A CENTURY OF FLIGHT

A CENTURY OF
FLIGHT

BILL GUNSTON

Brian Trodd Publishing House

Published in 1988 by
Brian Trodd Publishing House Ltd
27 Swinton Street, London WC1

ISBN 1 85361 001 1

Printed in Italy

CONTENTS

INTRODUCTION

For at least a thousand generations men have dreamed of flight. It must have seemed like an utterly impossible dream – an idea as ridiculous as flying to the Moon. Men did finally fly to the Moon in July 1969. However, modern technology is so powerful we might be forgiven for thinking that there is nothing we cannot do. Then we may be reminded that in 1895 a great savant solemnly pronounced 'There are no inventions left to discover'.

Plenty of people who were alive when that was said are still alive today. Within their lifetime falls the entire story of the human conquest of the air by winged flight. When early visionaries dreamed of flying they dreamed of winged flight, like the birds. But in fact humans first voyaged through the sky under balloons. Because balloons need no engine they were possible with 18th century technology. Some aeronauts ineffectually tried rowing a balloon like a boat, but to fly in the true sense the dirigible airship and then the aeroplane had to be invented.

It is only about 60 years since passengers began to fly from one place to another. They were bidden to wear their stoutest coat, heavy gloves and a leather helmet (the airline often provided the latter), and bring a woollen rug. Most airlines offered a stiff tot of cognac before boarding. Not a few passengers soon wished they had also brought pliers, wire, a soldering iron and a few other essentials to effect running repairs. Today they might be surprised to find that the frailest centenarian can cross the Atlantic in safety and comfort, at a speed faster than a rifle bullet or the Earth's rotation.

Unfortunately a book that tells the story of a century of flight must devote roughly half its content to war machines. During World War I Orville Wright, who started it all, said 'What a dream it was! What a nightmare it has become. . .' The pessimist of today could say 'Orville, you ain't seen nuthin' yet'. But hopefully this book indicates that there are also grounds for optimism.

Bill Gunston

A British Aerospace Buccaneer S2, one of 65 currently in service with the Royal Air Force.

1. THE VISIONARIES

THOUGH THE AVIATION INDUSTRY was born in the present century, we know from their writings and drawings that our ancestors dreamed of flying hundreds of years ago. Indeed, the glorious idea of swooping through the sky like today's Superman must have occurred to human beings tens of thousands of years ago.

How could this seemingly impossible dream be realised? Why, the obvious thing to do was make large wings like those of a bird and flap them vigorously. But this is easier said than done. There are problems of design and construction of such wings which are obvious to modern aircraft engineers, and we also know today that humans are simply not built to fly like this. We are too heavy, and much too weak – though in very recent years we have learned how to build extremely large yet light flying machines powered solely by their pilots. Hundreds of years ago the technology, the materials and the knowledge to build man-powered aircraft did not exist, and many would-be aviators tried to make up for this by foolhardy courage. They leaped off high buildings, wings flapping furiously, and were lucky if they only broke a leg.

It so happened that men first learned how to fly in a totally different manner, quite unlike the birds or any other of man's familiar flying neighbours. Because we are surrounded by our atmosphere, we tend to forget that air is a very material substance, and it weighs quite a lot if there is enough of it. A cubic yard weighs well over 2½ lb (an exact figure at sea level is 1.2255 kg per cubic metre), so if we could make something very large and airtight and filled with something lighter than air we could make it float in the atmosphere in exactly the way a submarine floats at different levels in the ocean.

In 1670 a Jesuit priest, Francesco de Lana Terzi, proposed a great aerial chariot lifted by four huge spheres from which the air had been sucked out. His idea was absolutely sound except for one small detail. The pressure of our atmosphere is really very great; if we were to add up the force of the air pressing on our own bodies it would come to a weight of over 20 tons (1,016 kg), about as much as 20 family cars. We do not notice this enormous force because it is exactly balanced by our internal pressures. But if we had a vacuum inside we should instantly shrivel under the crushing external forces. If de Lana had made his spheres light enough to be buoyant they would have been squashed flat by the atmosphere. Strong enough to withstand atmospheric pressure, they would have weighed many tons each. Even today we could not build a de Lana style 'aerial chariot'.

In 1709, Bartolomeo de Guzmao in Portugal is alleged actually to have constructed a small hot-air balloon, which was flown in the royal palace but had to be knocked down when it threatened to set fire to some curtains. It sounds a convincing story, but real evidence is lacking. For the dawn of undisputed human flight we have to come forward to 1782. By this time man had a choice of two lightweight fillings for a balloon. One was a newly discovered gas, popularly called inflammable air but today known as hydrogen. The other was at the time thought of as another special lightweight gas, but we now know that it was merely hot air. If any gas is heated and its pressure remains unchanged its density falls. Thus a balloon filled with hot air weighs much less than one filled with air at the temperature of the surrounding atmosphere, and it will be thrust up like a cork in water, finally stabilizing at some height above the ground at which the atmospheric density is the same as that of the hot air in the balloon.

None of this was known to the Montgolfier brothers, Etienne and Joseph, as they toiled at the family paper mill in Annonay, a small town in France some 50 miles (80.5 km) south of Lyons. There are many fanciful stories of how they got interested in the notion of flying. One says Madame Montgolfier's chemise (or, in another version, Joseph's shirt) was placed before the fire to dry and, filling with hot air, rose into the air. The more likely tale is that the brothers simply asked, "Why do the charred bits of waste paper rise into the sky above a

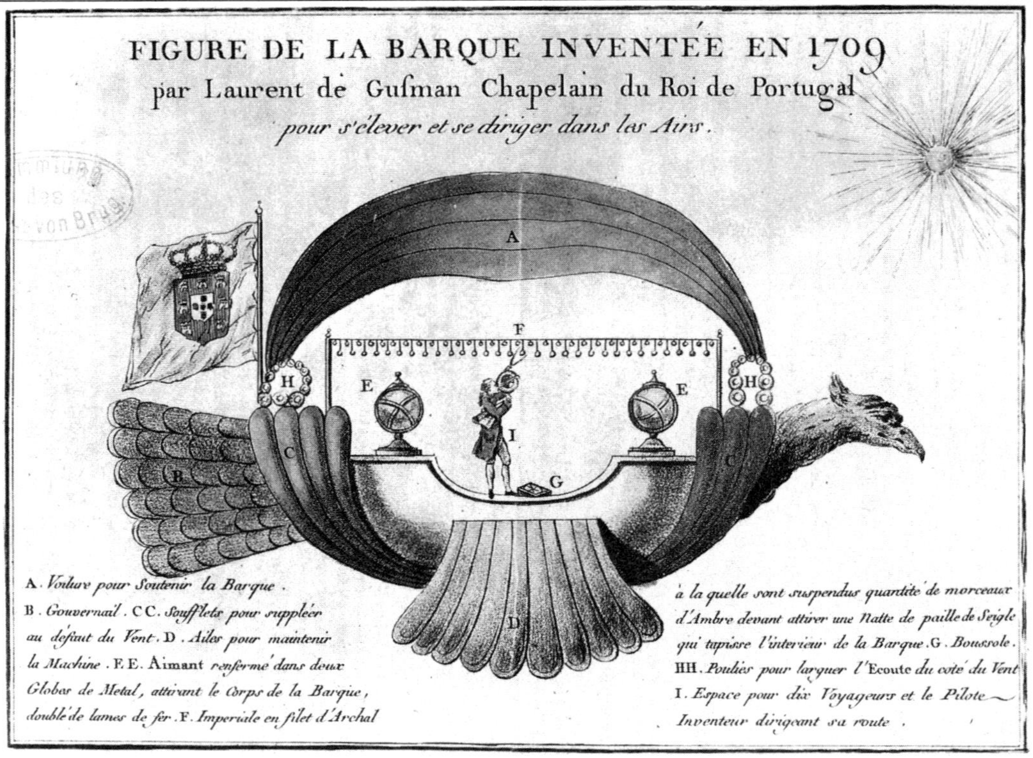

Above: Bartolomeo de Guzmao, here called 'Laurent de Gussman', never actually proposed this fanciful bird-like flying machine.

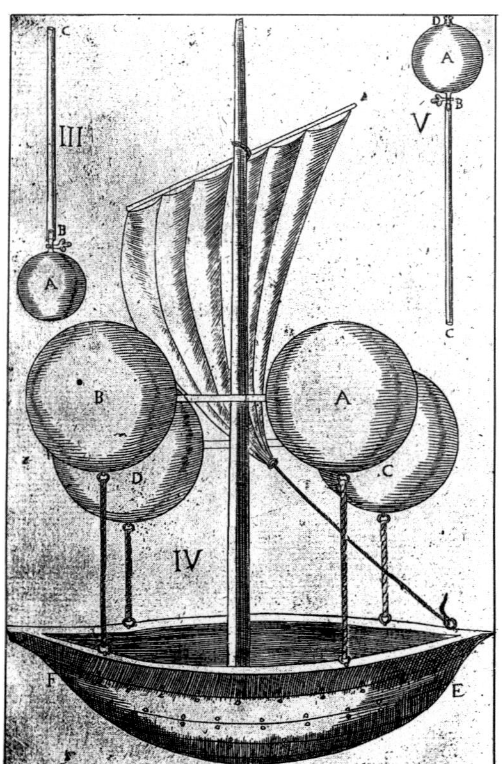

Left: Over 300 years ago an Italian Jesuit priest proposed building this 'flying ship' lifted by four spheres of copper from which the air had been sucked out.

bonfire?"

After making some small paper balloons, Etienne Montgolfier made an airtight bag of fine silk with a capacity of about 40 cubic feet. He tested this at Avignon in November 1782 and found that after holding burning paper under its open lower end for a few minutes the whole balloon took off. The next thing was to make a very much larger balloon, and on 19 September 1783 the Montgolfiers demonstrated a giant balloon before Louis XVI and his court. The passengers were a sheep, a cock and a duck, and they landed safely about 1⅓ miles downwind. Events now moved fast. With a man on board it would be possible to carry a bonfire along underneath the open end of the balloon and fly further.

The first aviator was Pilâtre de Rozier. On 15 October 1783 he boarded a colossal *Montgolfière* about 48 ft (14.6 m) in diameter and 74 ft (22.5 m) high, made of paper-lined linen and gaily painted. The paint must have added appreciably to the weight, estimated at 1,700 lb (771 kg). Around the base was a circular gallery from which the 'intrepid aeronaut' could throw dry straw bundles on to a brazier burning in the 16-ft (4.9-m) gap in the centre. On this first occasion the balloon was tethered, but it rose until the ropes were pulled tight. The stage was thus set for the first great date in man's conquest of the sky: 21 November 1783. At almost 2 pm de Rozier stepped aboard the same balloon, accompanied by a cavalry officer, the Marquis d'Arlandes, who balanced the weight on the other side. Casting off, they rose to about 280 ft (85 m) and swept off their hats to the mighty crowd below. They then threw on more fuel and, to quote the account of Tiberius Cavallo (himself a balloon experimenter), "they rose so high that the machine itself was scarce perceivable". They landed gently over 20 minutes later some five miles downwind. This was the first recorded human flight.

The Montgolfiers did not know their braziers merely heated the air; they thought the balloon rose because of some special gas given off, and spent much time searching for fuels that would make more of it. More scientific were Professor J. A. C. Charles and the brothers Robert, who were experimenting with spherical balloons filled with hydrogen. They made a manned ascent on 1 December 1783, watched by 200,000 Parisians.

Subsequently aeronauts – balloonists – sprang up in many countries. On 7 January 1785 the most famous of them, Jean-Pierre Blanchard, accompanied by an American, Dr Jeffries from Boston, set forth from beside Dover Castle to sail to France in a hydrogen

Below: An accurate portrayal of the lift-off of the full-size Montgolfier balloon, carrying a sheep, a cock and a duck in a wicker basket. The date was 1783.

balloon. Like most aeronauts of the day, Blanchard did not appreciate that because the balloon moves along with the local mass of air, such things as sails and rudders are useless. After throwing out almost everything removable, including most of their clothes, the two pioneer international air travellers reached France safely. But the fact remains to this day that balloons go where the wind takes them, and – notwithstanding strenuous early efforts to guide them with oars, rowed like a boat – they are fun items rather than practical vehicles.

Of course, it was later proved that a balloon can be turned into a *dirigible* (French for steerable) by adding an engine and propeller. We call such machines airships, and the earliest examples were so slow that they were useless except on a very calm day. The first to fly was Henri Giffard's steam-driven ship which first flew on 24 September 1852. But by far the most famous airship pioneer was Ferdinand, Count Zeppelin, a former Prussian general who was nearly 50 when he became convinced that the future lay with the large rigid airship, with a strong framework of aluminium enclosing its lifting gasbags. In 1900 his first 'Zeppelin' made its maiden flight, looking like a gigantic cigar. Zeppelins carried 10,197 fare-paying passengers in 1912–14, bombed many British towns in 1915–18 and made transatlantic and round-the-world flights in the 1930s, but when the biggest of them, LZ-129 *Hindenburg*, crashed in flames on 6 May 1937 it spelt the end of the era of giant airships.

This would have seemed not only tragic but also incredible to Count Zeppelin back in 1900, because at that time no powered flying machine of any other kind existed. Though the nineteenth century had been punctuated by numerous ideas for heavier-than-air machines – called aeroplanes in Britain and airplanes in North America – not one had shown that it could fly.

Prior to 1782 hardly anybody had given much thought to flying except using wings like a bird, so it is remarkable that at that time the science of winged flight was non-existent. There is no record of any experimenter making sets of model wings with small differences and testing them in a strong wind to see which gave the most lift, or even of working out the basic forces, weights, control problems and possible arrangements of a flying machine. But in 1799 a handsome Yorkshireman named

Right: One of several stylized pictures purporting to show the first human flight, by de Rozier and the Marquis d'Arlandes, in a Montgolfier balloon, 1783.

Left: The famous silver disc on which Sir George Cayley engraved his 1799 conception of a flying machine. It is considered the earliest known representation of an aeroplane.

Sir George Cayley arrived on the scene. He not only did all these things, so that he has ever since been called 'the father of the aeroplane', but much later, in the mid-nineteenth century, he even built gliders that carried a man.

In London's Science Museum is a small silver disc on which, in 1799, Cayley engraved the first-ever drawing of an aeroplane as we know it today, with fixed wings, a tail at the rear made up of a fin/rudder and an elevator (in the USA called respectively a vertical stabilizer and a stabilizer), separate propulsion provided by propellers behind the wing, and an underslung nacelle for the pilot. In 1804, the year before Nelson fell at Trafalgar, Cayley was busy testing modern-looking gliders and measuring the exact lift of wings mounted on a whirling arm! If only there had been a few more Cayleys, men might have made at least controllable gliding flights early in the nineteenth century. As it was, we only have records of numerous fanciful sketches and obviously impractical full-size flying machines that – fortunately for their inventors – did not fly.

What were the thoughts and objectives of would-be aviators in the late nineteenth century? Almost all of them concerned themselves with creating a flying machine. It would be incorrect to say that they were designers in the modern meaning of the word because it is impossible to design a flying machine without an underpinning basis of knowledge. Most of the budding flyers carried out no preliminary experiments or measurements, and seemed to think it quite natural to create a proposed aeroplane entirely in the mind. Even more incredible to us today was their supposition that, if only they could build it, they could simply step aboard and fly it.

Right up until about 1908, by which time many aeroplanes had actually flown, there were many constructors who continued to think that the task before them was to build the flying machine. The problem of learning how to fly – indeed, in most cases the need for the machine to have properly thought-out means of flight control – was totally overlooked. In recent years historians have called the dozens of members of this fraternity the 'aerial chauffeurs'. From 1900 it was common for wealthy motorists to employ a chauffeur to drive and look after their new motor car, and the would-be flyers rather took it for granted that, once they had built their machine, driving it through the sky would be very much like driving a car on the road. Thus, the failure of their often large and grotesque machines to rise from the ground saved them from injury or death.

There is little point in listing the profusion of names and ideas of which we have record, but a few were important in either coming near to flight or in influencing others. In Somerset, England, in the 1840s W. S. Henson and John Stringfellow built model monoplanes, with twin pusher propellers, which were sound in concept and lacked only suitable power – though Stringfellow built brilliant compact steam engines. He went on to exhibit a triplane at the world's first aero exhibition, held at London's Crystal Palace in 1868, which exerted a great influence on many other enthusiasts. In the same year a brave French sea-captain, J-M Le Bris, tried to fly in the second of two big gliders he based on the albatross. He made many short glides between 1857 and 1868 in his first machine and was lucky to break only a leg. Another Frenchman, Louis Mouillard, also studied birds and was one of the first to recognise that birds often glide, and that man might fly without trying to flap mechanical wings.

Yet another French inventor, Alphonse Pénaud, built small model monoplanes driven by a pusher propeller turned by twisted elastic. From 1871 he got these to fly, and not only were these made as toys by the million but they preceded the achievement of full-scale powered aeroplanes and led through to the rubber-powered 'duration' models of the 1930s, and on to the very similar rubber-powered models of today.

Three important pioneers who certainly got as far as sitting in the pilot's seat and opening the throttle were Mozhaiski of Russia, Ader of France and Maxim of England. Today's Soviet propagandists claim that Aleksandr F. Mozhaiski made the world's first powered flight, in 1884. Certainly he built a large rather square-winged monoplane with three steam-driven screws, which did rise just clear of the ground on being launched down a sloping ramp, but it had no means of control and certainly would have crashed had it got well clear of the ground.

The chief claim to fame of Clément Ader is that his first full-size machine, the *Eole* of 1890, did fly just clear of the ground for an estimated 164 ft (50 m). He was one of the 'chauffeurs' and his bat-like creation had no provision for any kind of flight control. Luckily for him, his bigger *Avion III* of 1897 failed to rise. In contrast, the giant biplane of wealthy Sir Hiram Maxim, inventor of the machine gun, did try to fly in 1894, but was restrained by a trolley fixed to a rail in Baldwyn's Park, Kent. Driven by a 180-hp steam engine of amazingly advanced and lightweight design, Maxim's monster did have primitive hinged control surfaces, but it was never allowed to fly.

Maxim combined a penetrating mind with immense financial resources, but there were other pioneers who were even greater in their ability to study the problem of the aeroplane scientifically. Three of the greatest were Lanchester, Phillips and Hargrave. Dr F. W. Lanchester, born near London in 1868, left a host of basic inventions used in modern cars, but he was also a great pioneer of aerodynamics, which is the science of air and other gases in motion. He discovered the boundary layer, the thin layer of air adjacent to a wing or other surface, which travels along with the aircraft and is usually turbulent. He coined the word 'aerofoil' for the cross-section shape of a lifting wing, and was the first to carry on where Cayley had left off in finding out the complex way that wings actually work.

Horatio Phillips, another Englishman, paralleled Lanchester in studying wings and discovering how they should best be designed. Back in 1884 he had patented the curved aerofoil with a round leading edge which dipped downwards, and he went on to build various full-scale aeroplanes which flew in circles tethered to a central post, mostly in Streatham or Harrow, near London. Hargrave had the disadvantage of being an Australian, cut off from any other people working in the

Right: Stringfellow's triplane model dominates this part of the 1868 exhibition held in the Crystal Palace by the two-year-old Aeronautical Society.

same field, yet he invented the box-kite in his search for a kite that would fly stably even in turbulent air. When he demonstrated it in Europe in 1899 he triggered off almost all the subsequent European aeroplane designers, as explained in the next chapter.

But there was one pioneer who, working entirely alone and doing everything for scientific reasons, spent far more time actually flying than everyone else in the nineteenth century combined. German Otto Lilienthal, born in 1848, was the exact opposite of the 'chauffeur'; he realised that, before he could fly an aeroplane, he had to learn how to fly. So he began with the simplest possible gliders, and from 1891 courageously made thousands of glides down steep slopes. In this way he not only learned how to make better gliders but he also learned how flying machines actually behave and how to control them. Unfortunately he concentrated on the method often called 'body-English', in which the machine is kept level merely by shifting the pilot's body and swinging his legs in particular directions. It would have been more useful if he had studied the effect of hinged control surfaces; his only hinge allowed the whole tail to rise freely, because he was afraid of a rising tail pushing the machine into a dive.

Lilienthal did not make a frontal attack on the powered aeroplane, though he did experiment with flapping 'multi-feather' wingtips driven by a carbonic-acid gas engine. Sadly, he died in 1896 after crashing heavily, probably because of the limitations of his control movements. His most immediate imitator was Percy Pilcher of England, who by 1899 had completed plans to build a modern-looking powered aeroplane. Had he been able to do so he would certainly have been able to fly it, but tragically he was killed in 1899 after crashing because of structural failure.

Above: John Stringfellow made what was probably the first model of a practical powered aeroplane in 1848. It was tested suspended from overhead wires.

Until the 1890s very little interest had been shown in flying machines in the United States, but an eminent civil engineer, Octave Chanute, emerged in the final decade of the century as the world's greatest patron of the early aeroplane, and an outstanding 'middle-man'. Though he did not fly himself, being well over 60, he made a classic survey of everything that had been written on the subject, and then sponsored a major attack on the problem of building a flying machine that would be inherently stable. He began with gliders derived from Lilienthal's, but ended the century with an outstanding biplane glider first flown in 1896 which was not only aerodynamically and structurally good but also had the pilot hung well beneath the wings on rails at each side. These rails enabled him to work his way fore and aft to balance the glider longitudinally, and thus prevent a catastrophic loss of control or a stall. Of course, as hardly anyone in the pre-1900 era had flown, the stall was unknown, though Lanchester and Phillips did discover it in their academic researches. Clearly the lift of a wing depends not only on its size and airspeed but also on the angle of attack (AOA), which is the angle at which it meets the oncoming air. At small positive AOAs the lift increases as the angle increases – so that, for example, ordinary aeroplanes have to fly increasingly nose-high as their speed is reduced, if they are not to lose height – but with most wings a limit is reached at an AOA of around 16°. Here the wing is only just able to keep the air flowing back across its arched upper surface, giving reduced pressure and thus lift. Any further attempt to increase AOA results in sudden breakaway of the flow; lift is reduced dramatically, and the aircraft virtually falls out of the sky, unless it has the ability to plunge nose-down, pick up speed and recover. Stalls killed thousands of early pilots, especially in World War I, even though by that time they were fully understood. Chanute was one of the first workers to try to avoid them.

Back in 1810 Englishman Thomas Walker had published a scholarly booklet on flying, proposing an ornithopter (flapping-wing machine). In 1831 he was back with a different idea: two fixed wings in tandem. Even today there is much to be said for this layout, which in theory can be made almost unstallable. In 1874 D. S. Brown picked up the Walker tandem-wing idea, and flew successful model gliders. This impressed an American, Samuel Pierpont Langley, a famed astronomer and Secretary of the Smithsonian Institution. In the 1880s Langley began testing small models, and his fifth model, in 1896, was an impressive machine with a steam engine. For better or worse, it flew with great success, showing natural stability.

The problem was that Langley was a 'chauffeur' and totally ignored the question of control. Even with a truly stable aeroplane, it reduces its usefulness if it is impossible to steer it, and landing would have been fraught with danger because it would have been on whatever happened to be in front. Despite this, Langley had the resources to build a full-size machine, and he also had the official 'clout' to obtain massive government funding. This was mainly because in 1898 war broke out between the USA and Spain, and it was suddenly realised in Washington that if an aeroplane could be built it might be a very useful military vehicle. Langley called his flying machines Aerodromes – this was long before the same word came to mean a flying field – and the 1898 contract with the US federal government was the first in history for a man-carrying aeroplane.

Throughout the nineteenth century, the flying machine was one of the pet subjects of newspaper cartoonists and other critics because it was almost universally regarded as something impossible of realization. People who tried to build or fly them were regarded as complete idiots, and fit only for ridicule. There was a perceptible change in the public attitude, however, when a person as brainy and famous as Langley got into the act, and especially with many thousands of US Treasury dollars behind him. Now at last, it was widely believed, Langley's Aerodrome would show the world that man could fly with wings.

Probably the best thing about the full-scale Aerodrome was its engine. This was an extremely advanced petrol engine with five cylinders arranged radially, like the spokes of a wheel. This is how almost all the most powerful piston engines of the 1940s and 1950s were arranged, and the Aerodrome engine even had cast-iron cylinders with steel liners, coil ignition and thin water cooling jackets. It was designed by Charles Manley, based on a car engine by Stephen Balzer. Manley was also the pilot of the Aerodrome, and he started off knowing that his remarkable engine put out about 52 hp for the incredibly low weight

Right: Twice S. P. Langley became a laughing stock when his carefully designed Aerodrome fouled the complex launch system and dived into the Potomac in 1903.

of 136 lb (61.7 kg) (not including the cooling water radiator). This power/weight ratio of 2.6 lb/hp was not to be reached again until well into the 1920s.

Unfortunately the Aerodrome was so big, with tandem wings each of 48 ft (14.6 m) span, that even this exceptional engine was inadequate. A second fault was that, like so many of his predecessors, Langley completely overlooked the need to provide some way of controlling the aircraft. As it happened, the most significant fault of all was the decision to launch the big machine by a catapult from the top of a houseboat on the River Potomac. It was 1903 before the Aerodrome was ready, and at last, in the presence of a mass of pressmen and others, the catapult was released on 7 October. The Aerodrome went off the end, pitched down and slid into the river – as one journalist put it, "like a handful of mortar". Repaired, the machine was put back on the catapult, and on 8 December 1903 courageous Manley revved up his engine for takeoff. This time unquestionably the frail structure fouled the complex catapult, and as the Aerodrome went off the end the front wing crumpled completely. Manley very nearly drowned.

A few people showed sympathy to Langley, though the US government lost all interest. Far more agreed with the editorial in the following day's *New York Times*, which crowed "The ridiculous fiasco which attended the attempt at aerial navigation by the Langley flying machine was not unexpected. The flying machine which will really fly might be evolved by the combined and continuous efforts of mathematicians and mechanicians in from one to ten million years – provided we can meanwhile eliminate such little drawbacks and embarrassments as the existing relation between weight and strength in inorganic materials".

The learned editor of that paper was actually somewhat wide of the mark. It was not to take "from one to ten million years" but precisely nine more days.

The Wright 1902 Glider in flight at Kittyhawk, North Carolina, piloted by Wilbur Wright.

2. THE PIONEERS

Of all the famous inventors of history there are few more universally known than the Wright Brothers. This is despite the fact that these two lanky young Americans were not only mild and retiring, but during the crucial years when they alone of all humans were getting out their 'Flyers' and droning around in the sky, hardly anyone knew what they were doing, and most attempts at publicity were met with derisive laughter.

As most of the world knows, the brothers' christian names were Wilbur (born 1867) and Orville (born 1871); there were other brothers and a sister who did not share the interest of the younger brothers in the flying machine, which was kindled when they were boys in Dayton, Ohio. Their first interest in business was in producing a local newspaper, but in 1892 they realised there was a boom in bicycles and opened a cycle shop which prospered. This enabled them to build up a background of capability in light engineering. Far more importantly, both brothers had what was clearly lacking in so many other early aerial experimenters: coldly logical and analytical minds, which thought through each problem, thought through each possible solution, and relied not on hearsay but on numerical values obtained by experiment. Admittedly, they read all they could about other would-be flyers, but, said Wilbur, "The one thing we eventually learned was utterly to discount all previous so-called knowledge". This was because, when the Wrights actually came to take their own measurements, they found them quite different from previously accepted values. Today we know what the Wrights at first found hard to believe, that they were right and all the previous experts had been wrong – often wildly wrong.

Of course, the Wrights themselves were not supernatural, and they too made some basic mistakes. One of the most fundamental was that, whereas almost all previous constructors of flying machines either sought to create an inherently stable machine or else never even considered the problem, the Wrights deliberately designed an unstable aeroplane and relied on their ability to control it. They also picked on a tail-first configuration that for the next 80 years was out of fashion – though today it is once more very competitive and in fact has much to commend it.

Wilbur and Orville were much intrigued by the rubber-driven toy aeroplane – it was a Pénaud-style fixed-wing toy, not a helicopter as often reported – given to them by their father, a bishop, in 1878. Ever after, they studied the aeroplane and its problems. One of the basic difficulties, ignored by virtually all the actual constructors of flying machines at the time, was lateral control. What happens, they asked, if a gust tips the machine up on one wing? Clearly the pilot must have a way of making the down-going wing give more lift and the high wing less lift than normal, to roll the wings back level. Around 1890 they were sketching arrangements for pivoted variable-incidence wings, so that the AOA could be adjusted by the pilot in flight – more on one wing, less on the other. They schemed an arrangement of gears at the centre which would make the wings pivot in opposite directions; but, clearly seeing how great would be the forces on the central pivots and how large the operating force would have to be, they decided there was no way the mechanism could be made light enough.

One day in 1899 Wilbur sold a customer a spare tyre inner tube. He kept the long, thin cardboard box in his hands. As both ends had been opened, the box was not rigid but easily twisted. Holding one end in each hand, Wilbur twisted the box so that one end was turned through almost 90° relative to the other. In a flash he pictured a pair of biplane wings, held together by linking struts in what is technically called a cellule. Could they not be made able to twist, like the box? If the pilot at the centre could pull strongly on wires or cords, could he twist or 'warp' the wings to roll the aircraft to left or right, thus providing the sought-after lateral control?

In August 1899 the brothers built their first large winged device, a biplane glider or kite of 5 ft (1.5 m) wingspan. The wings were linked by pivoted struts, so that the lower wing could not only change its angle but also simultaneously swing to the front or rear to preserve fore-and-aft balance. The glider was operated as a kite by two pairs of

Right: The cycle shop of the Wright brothers is today a national shrine in downtown Dayton, Ohio.

tethers, each connected to a short stick, so that the operator not only held the machine against the wind but could, by tilting the sticks, warp the wings and roll the machine. It worked beautifully.

Thoroughly committed, the brothers now knew they had to go on with machines they could fly themselves, ending up with a Flyer — as they called their early machines — fitted with an engine. While the design went ahead for a biplane glider of 18 ft (5.5 m) span, with two wings of 100 ft^2 (9.3 m^2) area each, and fitted with a front elevator to control it in pitch, they also considered where to fly it. Octave Chanute suggested the bleak sandy Outer Banks of North Carolina, a very narrow strip of land over 250 miles (402 km) long and separated from the mainland by many miles of sea. Kitty Hawk life-saving station replied to their letter about wind by saying the average was 15 mph (24 km/h), which sounded fine. The soft sand seemed just right if you were going to crash. But the journey in September 1900 took over a week, and in fact the boat part of it was highly dangerous. Wilbur went first, buying timber for the wing spars on the way. He wanted 18-ft (5.5-m) lengths of spruce, but the best he could do was 16-ft (4.9-m) beams of pine, which cut the wing area to only 165 ft^2 (15,3 m^2) total. He brought with him the coverings of French sateen, made up and sewn back in Dayton. Wilbur put up at the house of Kitty Hawk post-master Bill Tate, and used Mrs Tate's sewing machine to shorten the wing coverings.

Once Orville had arrived a few days later the complete glider took shape quickly. One major snag they discovered was that the 15-mph (24-km/h) average wind was typically made up of a calm one day followed by a 60-mph (96.5 km/h) gale on the next! Stinging sand whipped into their faces, and their big tent had to be secured by extra ropes, and firmly lashed to a tree. When they dared to get the glider out they discovered that, though all previous aeronautical tables told them it would lift in a 17-mph (27.4-km/h) wind, it failed to rise in a 24-mph (38.6-km/h) wind (the accepted figures were wrong). So they started off with the big glider loaded with 50 lb (22.7 kg) of chain instead of a pilot, and 'flew' it with cables from the front, as they had done with the small machine in Ohio. Gradually they gained confidence. After a few days they carried the machine four miles south to Kittyhawk Hill, a 100-ft (30.5-m) sand dune, and there they began gliding.

Above: The Wrights understood that they first had to learn how to fly. Here the No 3 Glider, modified with a rear rudder, is warping its wings to get the wings level at Kittyhawk in 1902.

The dune had a slope of 1 in 6, and keeping the machine at a steady height into a 15-mph (24-km/h) wind they found it gradually accelerated, so it did not need so steep a slope. The control proved to be all they had dreamed of. Warping the wings instantly corrected any tendency to roll, and the front elevator responded to the slightest movement of the pilot control. They had built the machine with dihedral, each wing sloping upwards from the centre-line to the tip, to get some natural lateral stability.

For the 1901 season at Kitty Hawk the Wrights built a much larger glider, with 22 ft (6.7 m) span and 290 ft^2 (27 m^2) of wing. They stuck to the front elevator, the only big change being to make the wings much more cambered (curved, or arched, from front to back), again according to 'accepted science'. Again, existing data were nonsense, and when they slightly flattened the wings they flew better. On the first flying day in 1901 they made a glide of 315 ft (96 m), but soon discovered that as their machine lacked 'weathercock' stability its wingtips moved first fast, then slowly. Warping the wings to lift one side also increased the drag on that side; the machine yawed (slewed round) so that the wing which was meant to rise moved so much more slowly through the air that it lost lift and, instead of rising, could even fall. For 1902 they needed a fin at the back.

It was in 1901–02 that the Wrights at last threw out every figure and report from previous experimenters, as being totally unreliable. Now they could really get somewhere, and their 1902 glider was a

fine machine of just over 32 ft (9.8 m) span with a double fixed fin at the rear. But there were still problems with warp drag more than cancelling the effect of the warping, and the fin often made the situation worse by swinging the up-going wing even faster, while frantic warping merely stalled the down-going wing. The final answer was to replace the fins by a single pivoted rudder and link this to the warp system so that, as the wings twisted, the rudder pulled the tail round to keep the aircraft flying in a straight line. The 1902 glider made over 1,000 flights, and by the end of that season both brothers were experienced pilots, the first in history.

The stage was thus set for the building of the first successful aeroplane. The Wrights wanted an engine of at least 8 hp, weighing without accessories not more than 160 lb (72.5 kg), but after many enquiries they decided to design and build their own. Assisted by their mechanic Charlie Taylor they made a four-stroke car-type engine with the water-cooled cylinders lying in a row on their sides, and with a number of features which today look odd, such as automatic spring-loaded inlet valves and with the gasoline (petrol) dripping by gravity into a hot well where it vaporised before being sucked into the cylinders. At 1,200 rpm the power was initially about 16 hp, but after less than a minute this had fallen to a steady level around 12 hp, which

Below: This famous photograph shows the start of man's first controlled flight in an aeroplane, December 1903. Orville pilots, Wilbur runs alongside.

was more than the brothers thought they needed yet about half what the engine should have produced. Bare engine weight was 152 lb (69 kg).

The aeroplane to be powered by it, later called Flyer I, was an enlarged version of the 1902 glider, and like the latter in its modified 1903 form had a double rear rudder. A new feature was a biplane front elevator. The engine was mounted on the lower wing on the right of the centreline, the pilot lying prone as in the gliders on the left, supported on a hip cradle which, pushed to the left or right on transverse runners, warped the wings to control the machine in roll. As before, the same system of cords also pulled on the rudders, to keep the Flyer on a straight course. The elevators were controlled by a stick held in the pilot's left hand. The engine had no throttle; once it had been steadied at full power for takeoff the pilot's only control was a string leading to the fuel cock to cut off the supply and, some seconds later, stop the engine.

The brothers were forever arguing, often each coming round to the other's viewpoint, and in so doing talking through each new problem. After a mammoth series of such discussions they decided to make their own propellers and use two of them. Going back to first principles, they designed propellers of outstanding efficiency (about 66 per cent), far superior to the crude screws with bent flat-plate blades used by many later pioneers up to 1910. The engine crankshaft drove a flywheel, to even out the speed, and two gearwheels, each meshing with a motorcycle-type chain. The chains drove horizontal propeller shafts held high above the lower wing on the struts, and the propellers were mounted on the rear ends of the shafts as pushers. One chain was twisted into a figure-8, running in tubes to avoid friction, so that the propellers rotated in opposite directions to avoid any gyroscopic action of the large spinning masses. Each propeller gave about 100 lb (45.4 kg) thrust at 302 rpm.

After more argument they did not fit wheels to the Flyer, but mounted it on skids like the gliders. First they laid out a 60-ft (18-m) sectioned wooden rail, as nearly as possible into wind. On this they rested a frame, running on two small tandem wheels, and then they placed the Flyer on the frame. Previously, the gliders had been

Right: On its first test the Wright Flyer I, which was an enlarged version of the 1902 glider, took off with Wilbur piloting, on 14 December 1903.
Problems with porpoising soon caused the plane to crash.

Left: Robert Esnault-Pelterie taking off from Buc airfield in June 1908. His REP No 2 was in most respects like a modern monoplane.

Left: Inside the Voisin brothers' aircraft factory in Paris. The machine in the foreground is the Farman I after reducing its tail span.

launched into wind with a helper on each wingtip, without any 'landing gear' at all.

As completed, Flyer I had a span of 40 ft 4 in (12.3 m), a wing area of 510 ft^2 (47.4 m^2) and a weight of some 605 lb (274.5 kg). With hindsight we can see that it was rather larger than necessary, even by the standards of the time, so that it was underpowered and difficult to control especially on a windy day. A more important fault was its basic instability, so that the pilot could never let up for an instant but had to fly it all the time. On the other hand, it was to be some time before flights were long enough for this to be a major problem.

All was ready on Monday, 14 December 1903. There was a gentle breeze, and the Flyer was at last free of snags, the last fault – slipping of the propeller sprockets on their shafts – having at last been cured by using the strong adhesive normally employed to stick cycle tyres to the rims! Wilbur won the toss, climbed aboard and tried to release the restraining rope. The thrust of the screws was too great, and several bystanders were needed to push the Flyer back to slacken the anchore. Orville, on one wingtip, soon found he could not run fast enough. As the Flyer went off the end of the rail he started the stopwatch. The Flyer climbed steeply to about 15 ft (4.5 m), lost headway and mushed heavily back into the sand, slewing round and breaking various skids and spars. At $2\frac{1}{2}$ seconds, this was not claimed as a flight.

In fact, the biplane elevators were pivoted too near the centre, so that they overcontrolled, first fully nose-up, then fully nose-down, but this was not recognised at the time in 'Will's' haste to blame himself. After repairs the Flyer was ready on Thursday, 17 December, and it was 'Orv's' turn. This time the wind was a strong 25 mph (40 km/h), but the brothers sensed history was in the making, because they rigged up their cumbersome camera and asked John T. Daniels, one of the men from the life-saving station, to work it as the Flyer left the rail. The result is one of the most important photographs ever taken. Pin-sharp, it shows the start of the first undisputed human aeroplane flight, with Wilbur having come to the end of his run and staring enthralled. In the foreground the disturbed sand outlines where the brothers had kept walking around the Flyer before the start.

Orville immediately found the pitch control grossly oversensitive, and proceeded in a series of switchbacks. On one of these he hit the ground, ending the flight, and the time was put at 12 seconds (the watch was not stopped immediately) and the distance 120 ft (36.6 m), which allowing for the strong wind was equivalent to over 500 ft (152 m) through the air. The broken engine shut-off lever was repaired, and then Will made a flight of 175 ft (53 m), the wind having slightly abated. Orv followed with 200 ft (61 m) in 15 seconds, and Will rounded off a marvellous morning with 852 ft (260 m) across the ground – well over half a mile air distance – in 59 seconds. This flight also was suddenly ended by the resumption of pitch-overcontrol, one of the undulations resulting in contact with the beach. Despite this, the brothers felt confident after lunch of flying the four miles to the life-saving station; but it was not to be, as the Flyer was overturned by a sudden gust and severely damaged.

This famous machine, the first aeroplane to fly, was later repaired, though it never flew again. Because of American refusal to recognise the Wrights' priority over Langley, in 1928 Orville (Wilbur died of typhoid in 1912) had the Flyer I shipped to London's Science Museum. In 1948 it was sent back, and it is today in Washington's marvellous National Air and Space Museum.

Orville sent a telegram to Bishop Wright at Dayton telling him of the flights and ending "Inform Press". But the newspapers were not only utterly skeptical but got the facts so confused their reports were almost meaningless, and not the least remarkable thing about man's first flights is that, apart from the Wright family and a tiny handful of others, nobody knew anything about them. This had an influence on the brothers, who increasingly became loners, doing their best to fly better but unconcerned about whether the world knew or not. They did, however, patent the 'Flying Machine', the award being granted three years after the 1903 application.

In the winter 1903–04 the Wrights constructed a much better engine, fitted it into their refined Flyer II and took it to a horse and cow pasture eight miles east of Dayton called Huffman Prairie. They flew here, again beset by many difficulties, about 80 times in 1904, the longest flight being 5 minutes 4 seconds. On 7 September 1904 they used a new invention, a takeoff catapult worked by a falling weight. This enabled flights to be made in a dead calm, but the brothers persisted with the idea too long and even in 1908 still used it, when they should have changed to normal engine-only takeoff with wheeled landing gears.

By 1904 many experimenters in Europe were close to flying, but for continuity it is best to remain with the Wrights, whose Flyer III of 1905 was certainly the first really good aeroplane. Today on display at Carillon Park, Dayton, Flyer III was much more robust than its predecessors, had no major control problems and few shortcomings apart from the use of skids, a launch track and a falling weight for takeoff. On it the brothers learned to fly under full control, to make banked turns, circles, figures of eight, climbs and dives and then controlled landings. Though Flyer III only made 49 flights, between 23 June and 16 October 1905, these at last were really impressive. For example on 4 October Orville flew for 33 minutes 20 seconds, and the next day Wilbur stayed aloft 38 minutes 3 seconds, at a speed of some 38 mph (61 km/h). Still nothing got into the papers, except for an article in the French *l'Aérophile* of December 1905 which few of its European readers seemed to understand; they certainly did not comprehend that the Wrights had not only understood the problem of how to control an aeroplane but had solved it.

The Wrights did not fly between 16 October 1905 and 6 May 1908. They had done all they could for the moment, and felt they were now being pressured to "give up their secret". Instead they got on with their real business, and where flying machines were concerned sought a clear-cut business deal with a client, preferably the US or some other government. This proved difficult, and in retrospect it would have been far better had the Wrights simply set up a factory to build the Flyer III. They could have paralleled this with the first flying school. This really would have furthered aviation better than their increasing feeling that they were under siege.

While the methodical Wrights were conquering the air, hundreds of Europeans were individually pursuing their own often ill-thought-out conceptions. By the time of the Wrights' first real flight in 1903 possibly as many as 60 full-size aeroplanes had actually been made, all unsuccessful. About as many more were made in 1903–08, at the end of which period the Europeans, led by the French, were flying at last. In the first years of the century Ferber, Cody, Weiss, Archdeacon and Voisin were leaders in the glider fraternity, the favoured concept being the Hargrave-style boxkite. In June 1905 Gabriel Voisin's tandem boxkite glider was towed at a good height behind a speedboat down the Seine. In 1906 came the first tentative powered hops. Trajan Vuia, a Paris-based Romanian, just got his monoplane into the air but it lacked control. Danish Jacob Ellehammer built an 18-hp monoplane with a second loosely-braced sail-wing above, and made a sustained flight on 12 September 1906 – but round a circular track, tethered to a post. He too had no proper control and would have crashed if he had tried a free takeoff

Thus the palm of being the world's first aviator after the Wrights goes to tiny Brazilian Alberto Santos-Dumont, darling of Parisian society, who used to park his airships at his own front door in the Champs-Elysées, and in 1901 made a 7-mile (11.3-km) flight from St Cloud round the Eiffel Tower. He considered trying a helicopter, but wisely settled for an aeroplane, and the machine he constructed looks not unlike other early biplanes except that we find the pilot is standing facing the tail! In fact it was totally back to front, with the biplane tail at the nose and the superb Antoinette vee-8 engine and metal propeller pushing at the tail. On 23 October 1906 this machine, the No 14bis, made a hop of 200 ft (61 m). With ailerons added for lateral control, and the 24-hp engine replaced by a 50-hp version, the 14bis then made six brief flights on 12 November 1906, the longest covering 722 ft (220 m) in 21 seconds.

This brought Santos-Dumont even greater fame, but nobody copied his tail-first aeroplane. After 1907, he built and flew tiny monoplanes called Demoiselles, which by 1909 were quite practical machines, although rather tricky. But the mainstream of European development centred on the brothers Gabriel and Charles Voisin, whose machines were of the tandem boxkite variety, though copying the Wrights in having a biplane front elevator. Many had side curtains, vertical fabric walls joining the struts between the wings. The first to fly was Delagrange No 1, named for its purchaser – the Voisins being the world's first sellers of flying machines. It made six

Below: A standard production 1909 Voisin – actually No 16 off the assembly line – with forward elevator, nosewheel and side curtains.

brief hops in spring 1907, the longest being 200 ft (61 m) on 30 March. Modified, it flew 1,640 ft (500 m) in 40 seconds on 5 November 1907, but only four days later this was beaten by another Voisin built for Paris-based Englishman Henry Farman. He continually modified his machine and, after teaching himself how to handle it in brief hops from 30 September, progressed to a flight of almost a minute on 26 October 1907, a controlled turn on 8 November, and a flight of 3,380 ft (over 1 km) in 1 minute 14 seconds on 9 November. On 13 January 1908 Farman became the toast of Paris by flying the first officially observed circular flight, though in fact this was shorter than a previous straight trip.

On 5 April 1907 a new name came into the picture: Louis Blériot. A prosperous engineer, maker of all the best car headlamps, he had been trying to fly since Blériot-Voisin gliders of 1905, but thanks to Vuia he switched to powered monoplanes and on 5 April 1907 at last made a brief hop. With the Blériot VI he flew 500 ft (152 m) on 25 July 1907, and his flights grew ever longer and more purposeful with each new machine. Another French pioneer was Robert Esnault-Pelterie, who made aeroplanes and engines with the trademark REP. On 16 November 1907 his REP.1 monoplane – which seemed to be almost all-wing, with two of its wheels on the wingtips – made a good 55-second flight over a distance of 2,000 ft (609 m). Like most machines up to this time it used wing warping, but the use of hinged ailerons gradually emerged as a superior method. In fact, Esnault-Pelterie was the first person to fly with ailerons, on a 1904 glider that was otherwise based on the Wrights', but he completely failed to do any serious analytical or measured experiments, and like almost all the pioneers other than the Wrights he relied on hunches and dogmatic beliefs.

All this time the Wrights continued to be beset by spies, and by unreasonable demands by the US government, so that no Wright Flyers were sold and the brothers kept their further work carefully locked away from prying eyes. But in 1907 they began to build a series of ten more or less standard Flyers, called the Model A, with 30-hp engine, 410 ft² (47.4 m²) of wings and an empty weight of about 800 lb (363 kg). The greater power enabled a passenger to be carried, and though they retained the

Right: The first crash by a powered aeroplane injured Orville Wright and killed his passenger, Lt T. E. Selfridge, US Army (who had flown as a pilot). Ft Myer, 1908.

Left: Early aviators were often completely exposed to the elements, but seldom as much as in the Antoinette monoplanes. You sat on them rather than in them.

interlinked warp/rudder control (it varied slightly from machine to machine) the pilot and passenger were now able to sit upright on the lower wing, with their feet on a fixed rail.

During the Wrights' 2½ years hiatus they had tried to get the US War Department interested, pointing out that they did not seek any money but merely wished to be told what requirements an aeroplane would have to meet to be of use to the Army. The result was a series of letters which shows for all time that either the War Department were obtuse to the point of lunacy or that for some unknown reason they wished to misconstrue what the brothers wrote and be as obstructive as possible. The Wrights therefore entered into negotiations with a company in France, Charles R. Flint & Co., and in July 1907 crated up a new Flyer A and shipped it to Le Havre. It remained there while further negotiations took place, but at last contracts were signed with the US Army in February 1908 and with France in March. The brothers then decided to come out of hiding, Orville going to Fort Myer at Washington for the Army tests and Wilbur to Europe.

First they had to regain their piloting skill, and so they took Flyer III of 1905, modified to carry a pilot and passenger both seated upright, and returned to Kitty Hawk. They were accompanied by their mechanic Charles W. Furnas. After putting right a lot of damage to their camp from storms, and from locals who thought the brothers would never return, they put in intensive flying practice from 6 to 14 May 1908. On the final day each brother gave Furnas a ride,

and he thus became the world's first aeroplane passenger. These flights were witnessed by several lurking journalists, unknown to the Wrights, but many of the readers refused to believe the stories they filed – which were the first reports to give any kind of true account of what the Wrights had been doing for five years. Indeed only a short time before the Paris *New York Herald* had commented "They are in fact either flyers or liars . . . It is difficult to fly; it is easy to say 'we have flown'."

All these years of aggravation and disbelief were about to end, but before the Wrights at last received their due recognition somebody else actually flew in public in the USA, and stole a lot of the Wrights' thunder. In September 1907 a group of Americans led by the famous inventor Alexander Graham Bell, and financed by Mrs Bell, met at Hammondsport, in western New York state, and formed the AEA (Aerial Experiment Association). A leading member was motor and motorcycle engineer and racer Glenn Curtiss, and his excellent 40-hp aircooled engines were used in the first three AEA aircraft, all biplanes influenced by the Wrights and the Voisins.

The first AEA machine, US Army Lt T. E. Selfridge's *Red Wing*, made a 319-ft (97-m) hop on frozen Lake Keuka, piloted in its builder's absence by Canadian F. W. Baldwin, on 12 March 1908. The second, Baldwin's own *White Wing*, was flown by various AEA members on 18 to 23 May 1908, Curtiss flying it for over 1,000 ft (305 m) on 22 May. Curtiss himself built a quite complex machine, *June Bug*, which flew more than 30 times between 21 June and 31 August 1908, its best flight being over 2 miles (3 km) on 29 August. The irony is that on 4 July, appropriately Independence Day, this machine won the *Scientific American* trophy for the first officially observed flight longer than one kilometre in the USA! The Wrights had made a 24-mile (38.6-km) flight three years previously.

Wilbur Wright went to France in late May 1908, and spent eight weeks getting the giant crate through customs, transported to the Le Mans factory of his friend Leon Bolée and assembled for flight. Maddening problems persisted. On 4 July, the day on which Curtiss won the *Scientific American* prize, Wilbur was badly scalded when a water hose broke during engine ground tests. This inevitably caused a delay, and a Paris headline proclaimed LE BLUFF CONTINUE. Not until 8 August 1908 was all ready, with the Flyer A set up on its launch rail on the Les Hunaudières racecourse outside Le Mans. A big crowd gathered, and in it were hundreds of highly critical people, such as journalists and members of l'Aéro Club de France, most of whom were avidly explaining to each other why the 'boastful American' would soon be deflated, even after the Wright's engine had started.

Wilbur flew for only 1 minute 45 seconds on that occasion, but he left the crowd shocked, stunned, amazed and almost speechless. Seeing a flying machine was not new, but here for the first time they saw a machine that really flew. It performed positively controlled banked turns, this way and that, zoomed towards trees, pulled up and swept round past them, circuited the field and then came in to an accurately controlled landing. Famed French pilot Delagrange spoke for the whole of Europe as he simply said "Nous n'existons pas!" ("We do not exist").

Unquestionably this one brief flight revolutionized aviation. There were plenty of

Below: Latham sets out for his second attempt to fly the Channel, on 27 July 1909. Minutes later his Antoinette had to ditch.

things to criticise in the 1908 Flyers, and by World War I this family of aircraft was obsolete, but that one short demonstration opened the eyes of the Europeans to what flying was all about. Any ideas of 'chauffeurs' were forgotten. Suddenly it was appreciated that the aeroplane moves in three dimensions, can dive, climb, yaw and roll, and that the man at the helm has to learn the totally new skill of piloting. Merely 'getting daylight under the wheels' was no longer enough. What was also suddenly obvious to everyone was that, to have such complete mastery of so refined a machine, the Wrights must indeed have been flying back in 1903.

Subsequently Wilbur moved a few miles to a big French Army artillery ground at Auvours. Here he rounded off the 1908 'season' with a string of really great flights, including 60 with passengers on board. There were six trips of greater than one-hour duration, and on the last day of the year he flew 78 miles (125.5 km) in 2 hours 20 minutes.

Back in Washington Orville's Flyer was checked over by the Signal Corps, and flying then began on 3 September 1908. His demonstrations were as spirited as his brother's, and though he made only ten trips, five were of about one-hour duration, and on three he carried passengers. Tragically, the series was cut short by a fatal crash on 17 September. One propeller blade cracked, setting up a sequence of vibrations and further failures which caused an uncontrolled dive. The pilot got the nose partly up, but the impact was hard enough to fling out Orville, causing serious injuries (he re-

covered rapidly). His passenger, none other than the AEA's Lt Selfridge, was killed, the first person to be killed in an aeroplane. Fortunately, the USA, and the Army, were thrilled at Orville's preceding demonstrations, and could hardly wait for him to recover and carry on.

In 1909 the Wrights were fêted everywhere, and production Flyers began operating in Europe and with the US Army Signal Corps. In England the balloon firm of Short Brothers not only obtained a licence to build Flyers but in fact produced the first set of factory-type working drawings, which the Wrights had never needed!

But 1909 was famed for two events both based in France, which continued to be the place where new-technology people gravitated until 1914. The first was one of the truly great individual flights of history. The biggest British national daily newspaper, the *Daily Mail*, owned by dynamic Lord Northcliffe, became a great champion of aviation and was to offer many big prizes in its cause. It offered £1,000, today worth perhaps 100 times as much, for a flight across the English Channel. The leading contender was Hubert Latham, who flew one of the most distinctive and beautiful machines of the day, an Antoinette. Technically much more advanced than a Wright flyer, it looked almost like a modern monoplane, with all the component parts in the places modern eyes expect to find them, and with the tractor propeller turned by one of

Below: All the A.E.A. machines looked very much alike. This was the one built and flown by Curtiss, the 'June Bug'.

the superb Antoinette engines. By sheer bad luck the engine for once let Latham down, and he had to ditch in the Channel on 19 July 1909.

It was left to Blériot to have a go, despite a badly-burned left leg caused by an in-flight fire that might easily have been catastrophic. Early on the morning of Sunday 25 July he asked a bystander "Which way is England?", got aboard his Blériot XI(mod) and took off. At first all was well, but about 12 minutes later the aviator was utterly alone in a grey mass. Though not in cloud, he could see no horizon, nor any sign of land. More than ten more minutes elapsed before Blériot suddenly found himself almost on the famed white cliffs of Dover. The wind was so strong he could hardly beat a path across them, but he found a gap, with the green meadow of Dover Castle beyond, and eventually managed to land – rather heavily.

Immediately the face with the big black moustache was in every newspaper, and Blériot became not only the 1909 equivalent of today's pop stars but also the most prosperous planemaker of the next several years. His neat Type XI monoplanes were tough, fast, agile and cheap, and they were among the few pre-1910 aeroplanes to play a significant role in World War I.

The second big 1909 event came only a few days later. In order to publicise champagne and its region, the city of Rheims decided to organise the world's first big aviation meeting. Large cash prizes were

Below: J. T. C. Moore-Brabazon preparing for flight in this new Voisin, at Issy on 1 December 1908. He was awarded the Royal Aero Club Aviator's Certificate No 1 (inset).

Fédération Aéronautique Internationale.
British Empire.

We the undersigned recognised by the F. A. I. as the sporting authority in the British Empire certify that | Nous soussignés pouvoir sportif reconnu par la F. A. I. pour l'Empire Britannique certifions que

J. T. C. Moore-Brabazon

Born at London on the 8 Feb. 1884,

having fulfilled all the conditions stipulated by the F. A. I. has been granted an | ayant rempli toutes les conditions imposées par la F. A. I. a été breveté

AVIATOR'S CERTIFICATE. PILOTE - AVIATEUR.

THE ROYAL AERO CLUB OF THE UNITED KINGDOM.

Roger W. Wallace Chairman.

Harold E. Perrin Secretary

Date 8 Mar. 1910 No 1.

J. T. C. Moore-Brabazon
(Signature of Holder)

donated by the leading champagne firms, and the wonder is that only 22 pilots took part. Some shared aeroplanes with others while a few were able to fly several machines, of which 38 were entered. In all respects, the Rheims meeting, held on the beautifully-arranged fields at Bétheny just outside the city, was unlike anything the world had ever seen.

A *piste* (racetrack) 6¼ miles (10 km) long had been prepared, with a gigantic grandstand, 600-seat restaurant where champagne flowed like water and haute cuisine was of the highest standards, and everything the aviators, mechanics and the Press could desire. People came in their thousands, including top politicians, generals, admirals and *hommes des affaires* (businessmen). At last they could see what this aviation was all about. Could aeroplanes carry people or cargo? Could they be useful for reconnaissance? Could they drop bombs or carry guns? What did they cost, and which were the best?

Answers to the latter question varied. Certainly young Curtiss from America suddenly came into the limelight, because thanks to his powerful engine he carried off both the top speed prizes, with speeds of 47 mph (75.6 km/h) over two laps (Coupe Gordon Bennett) and 46⅔ mph (75 km/h) over three laps a day later (Prix de la Vitesse). Winner of the biggest single prize, of 50,000 francs, was Henry Farman, who flew the greatest non-stop distance for the Grand Prix de la Champagne et de la Ville de Rheims. He covered 112 miles (180 km) in 3 hours 5 minutes, but in fact did not land for another 10 minutes after that. Farman also won the Prix des Passagers for making a lap with two passengers on board at once. Latham took the Prix de l'Altitude at 509 ft (155 m), though in fact both he and other pilots had gone much higher on earlier occasions. The prize for the fastest speed over one lap was won by Blériot, at 47¾ mph (76.8 km/h).

The Rheims meeting served to show the world that, though still an infant, aviation was a lusty one. Few noticed the absence of participants from Germany, yet it was in this country that the aeroplane was most immediately to become a production weapon. Only three years after the Rheims meeting, British car engineer Roy Fedden – later to design all the famous Bristol piston aero engines – went to the Mercedes company at Stuttgart to talk about electric car lighting. By mistake he wandered into the wrong part of the factory. He was surprised to find lines of big engines, all newly-built. With a thrill almost of horror he suddenly realised that they were engines for flying machines. Such a sight was conspicuously absent from Britain, and it seemed to bode ill for the future.

The Caudron GIII, which first flew in May 1914, was used during World War I initially on reconnaissance and artillery spotting and later as a trainer. The one pictured is a replica owned by the Jean Salis Collection and built by its founder; it has made several Channel crossings.

3. THE KNIGHTS TAKE TO THE AIR

Above: A rare bird indeed, the Royal Aircraft Factory B.E.3, with Royal Flying Corps (Army) number 203. It was a rotary-engined development of the B.E.2.

THE IMPRESSIVE PRODUCTION LINE of Mercedes engines at Stuttgart in 1912 did not happen by chance, or because that company had any particular faith in aviation. It was because they were ordered under military contracts and paid for. Back in 1909 the German government in Berlin had voted the equivalent of £40,000 to build up army aviation, and this was quite apart from another large sum for naval airpower based chiefly on Zeppelin airships.

This soon became known, and spurred off a somewhat lesser effort by the French, who had the advantage of almost three-quarters of the world's existing pilots and aerodromes. But in Britain the government, forever economy-minded, realised with concern that it had already spent £2,500 on poorly managed aeroplane experiments, and promptly stopped any further expenditure. Perhaps more than in any other country, the official view in Whitehall and in the British Army was that flying of any kind other than ballooning was for lunatics and to be ignored. Balloons had such a long history that they could be accepted, and in 1908 the Balloon Factory at Farnborough was graced with the title 'His Majesty's'.

The CO of that establishment, Lt-Col J. E. Capper, was an officer of great vision. Paying his own fares, he visited the Wrights at Dayton and secured from them an offer to work exclusively for the British War Office for four years (unless, as looked unlikely, the US government requested their help), for a sum of £20,000. Capper urged the British to accept, but lost his case! The prevailing view was that the horse and the rifle were adequate, and even the machine gun was a very doubtful asset. Aeroplanes were pointless.

In this view the generals and ministers were to a large extent reflecting popular opinion, though this was certainly modified by Blériot's cross-Channel flight. To a considerable degree the coming of the aeroplane split families into the parents, and especially the patriarchal father whose word was law, and who loftily decreed that the aeroplane was a piece of nonsense fit only for rash fools (as they had been saying for years about motor cars), and the younger generation whose outlook was more open-minded. Inevitably, the top military men thought like the older generation, and a junior officer who showed any enthusiasm for aviation might damage his career.

Military aviation in Britain began with two pioneers. One was the flamboyant and publicity-minded S. F. 'Buffalo Bill' Cody. He was an American who constructed and flew man-lifting kites for the Army at the turn of the century, and thus had the inside track when he built and flew British Army Aeroplane No 1 at Farnborough. It made a hop on 16 May 1908 but, as it was taxiing along on the ground, it hit a horse trough! On 16 October 1908 it flew 1,390 ft (424 m), and this is regarded as the first aeroplane flight in Britain. The other pioneer was a serving officer, Lt J. W. Dunne, who built aircraft with swept-back wings modelled on the winged seeds of the Zanonia plant. He tested his machines secretly on a private estate in Scotland, and did much to educate the generals; he even persuaded the government to appoint a committee to think about aeroplanes. An Air Battalion was finally formed on 1 April 1911, and on 13 May 1912 the Royal Flying Corps was formed. By comparison with some other countries, notably Germany, it was puny. Almost all of its officers had learned to fly at their own expense and its equipment was a rag-bag of mostly outdated machines.

The British government got off on the wrong foot with British manufacturers from the start. It gave them no help and virtually no orders, so that no good aeroplanes emerged. Then, in 1912, when the RFC got going, government orders went mainly to France. Col Seely, Under-Secretary of State for War said "We cannot buy British machines at the cost of human life", omitting to note that the fault lay wholly at the door of the government. Sadly, exactly the same poor relationship was to endure through the 1960s; more than any other nation, Britain was to fail to achieve the essential consistency in military procurement that is the basis of airpower.

But in 1912 the very concept of airpower was non-existent. It was left to individual officers, often with no direction or authority from above, to find out what the aeroplane could do. In the van in such experiments was the US Army which in February 1908, despite tight funds, had been the first military force to buy an aeroplane. In August 1910 one of the young Signal Corps officers, Lt Jake Fickel, who had been taught to fly a Wright biplane, went to an air meet at Sheepshead Bay. He rode as passenger in a Curtiss and succeeded in hitting a target on the ground with a rifle, despite the fact that the lower wing had no passenger seat, safety belt or hand-hold. A few days later Lts Crissy and Beck dared to make the world's first aerial bombs, using variously 3 in (7.6 cm) artillery shells and explosive-filled lengths of $2\frac{1}{2}$ in (6 cm) pipe, with a nose fuze and tailfins. They dropped them, well away from the crowds, on the Tanforan racetrack at San Francisco where another big air meet was being held, and achieved

Below: 1911. The first US Navy aviator, Lt Theodore G Ellyson, at the Controls of the Navy's first flying machine (Curtiss A-1).

Below: Here seen with a two- (instead of the usual four-) bladed propeller, the Royal Aircraft Factory B.E.2a was the first standardized production aircraft built in Britain. Powered by an imported Renault aircooled V-8 of 70 hp, it gave good service but was unable to manoeuvre quickly. At first its stability was judged a virtue, but by 1916 it meant that later versions could not dodge German bullets; they became 'Fokker fodder'.

100 per cent success!

Similar pranks took place in other countries, and by 1912 most military pilots were beginning to receive formal instruction in map-reading, reconnaissance observation and note-taking. Aerial photography was also taught, although this demanded the use of the very few aircraft that could carry a passenger, film plates and a camera. In those days this was an enormous box demanding eleven separate operations in order to take a single photograph, and with heavily-gloved or frozen hands this was no simple matter. Leaning out could be dangerous, because safety harness was not used by the European aviators and parachutes were discouraged. (To the end of World War I the official Allied view was that they might encourage cowards to abandon their aircraft.) Pilots learned to wear a leather helmet the hard way (US flyers received lifesaving crash helmets, but these were later discontinued) and, after a few had been hit in the eye by hailstones or beetles, goggles.

In parallel with these first steps in military aviation, an even more exclusive group was pioneering airpower at sea. Again, one of the first countries to take action was the USA, which picked Capt W. I. Chambers on 26 September 1910 as the officer 'to whom aviation correspondence should be addressed'. Not unnaturally he picked Curtiss as the Navy's first supplier of aircraft and also began organizing a course of instruction for navy aviators. A big step forward was the first takeoff of an aeroplane from a ship and the first landing on one – in both cases, at anchor. Both were accomplished by a Curtiss company pilot, Eugene Ely, the takeoff from the cruiser *Birmingham* in Hampton Roads (not far from Kitty Hawk) on 14 November 1910 and the landing on USS *Pennsylvania* off San Francisco on 18 January 1911. By this time Curtiss had also sold seaplanes to the Navy, and, landing beside a warship, had been hoisted aboard, enjoyed hospitality in the wardroom, and then been swung out over the side again and taken off.

The aeroplane unexpectedly found its first real use in warfare in the sandy desert of North Africa. In 1909 Italy and Turkey declared war, and the Italian army near Tripoli had a small group of aeroplanes, most of them French-built. On 22 October 1911 the first mission was flown by Capt Moizo, who brought his Nieuport down to

Above: 'Buffalo Bill' Cody won the British Military Aeroplane Trials on Salisbury Plain with an outdated aircraft; the secret was its 120-hp Austro-Daimler engine.

about 600 ft (183 m) in order to see more clearly what the Turkish troops were doing. Several rifle bullets passed through his aircraft, but Moizo was unharmed. Three days later he went out accompanied by Capt Piazza, who flew a Blériot. Their reports were of great value in preparing for the battle of Skiara-Skyat the next day, when Piazza and Lt Gavotti acted as artillery observers, directing Italian gunfire onto a battleship and the mountain artillery. Again they experienced fire from Turkish guns of many calibres, but returned safely.

An even more prophetic mission took place on 1 November 1911, when Lt Gavotti went out in his Etrich Taube on the first bombing attack. He had four small bombs carried in a leather bag in the cockpit and the detonators were in one of his own pockets! When he arrived over the main Turkish camp at Ain Zara he took out a detonator, inserted it into a bomb and released it. Despite a barrage of fire from the ground he did this several times, causing panic below. Perhaps the supposedly bold Turks were embarrassed at the effect caused by four bombs not much bigger than grenades, because the kicked up a great international fuss and claimed the Italians had committed an outrage by 'bombing a hospital'.

The Turks were again mainly on the receiving end of airpower in the Balkan war, which began in September 1912. The main opposing country was Bulgaria, which started with just two Blériots. It is worth noting that these were monoplanes, like all the Italian machines which did so well in the 1911 war; whereas in Britain an anti-monoplane feeling grew up in the armed forces which almost precluded the use of such aircraft until well into the 1930s!

In the Balkan war aircraft were used in a totally haphazard and unplanned way. Machines were bought wherever they could be found, and pilot training was such that five of the first six officers recruited in Russia were killed in five weeks. Eventually operations of a sort began in the Adrianople area and more than 300 bombs were dropped, but all but a few of these were locally made missiles based on shells, grenades or even tin-cans packed with explosive. They were released in every conceivable manner, one Russian pilot preferring to hang them from his feet and release them with a kick! The few properly made bombs were supplied from a factory in Italy which, spurred by the 1911 war, had got into production with the first aerial bombs ever designed.

These bombs had first been used on 10 March 1912 when Turkish positions were bombed by an Italian Forlanini airship. At this time aeroplanes were puny, and even getting into the air was a struggle, especially with a large pilot. Airships, however, were able to carry bombloads substantial by comparison, and both the German army

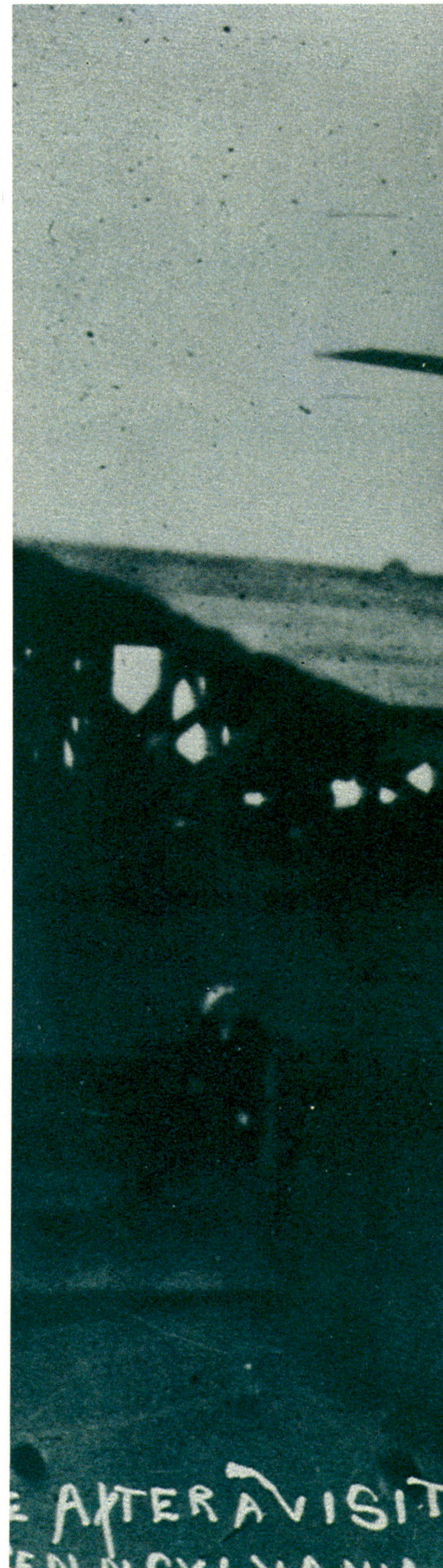

Right: On 18 January 1911 Curtiss pilot Eugene Ely made the first landing on a ship – the cruiser USS Pennsylvania *– and then, as seen here, took off again.*

BOARD

Above: In World War I millions of men had to be taught new skills in only a few weeks. Here a class of RFC pilot cadets learn the B.E. airframe from a corporal.

and navy looked on the airship as the chief new means of carrying firepower to a distant enemy. The army had mainly Schütte-Lanz ships, which had a wooden skeleton, and built up a force for use over land battles in such roles as reconnaissance, artillery spotting and bombing. The Navy used Zeppelins with an aluminium framework and practised long sea missions collaborating with surface fleets. Under Peter Strasser the naval ships also learned all-weather navigation and bombing, gradually perfecting their equipment until it was possible to navigate several hundred miles and make accurate landfall. Bombing accuracy from airships was poor by modern standards, typically achieving an error of about 1,200 ft (370 m) from 10,000 ft (3 km) altitude.

Even this performance was much better than would have been possible with aeroplanes, though of course pre-1914 flying machines could not get near such an altitude even with no bombs. The first properly conceived bombing apparatus was designed by Riley E. Scott, a former US Army officer, whose 'bomb dropper' comprised a frame carrying two bombs, a telescope and a table of figures for speed and altitude. Attached to a Wright of the Signal Corps in 1911 it prevented the machine from getting airborne except with a light pilot, but gradually accuracy improved. In 1912 Scott went to France where in a demonstration at Villacoublay, one of the world's first properly organised military airbases, he hit a 60 ft square 12 times out of 15 from 656 ft (200 m). American disinterest was total, but in Europe it was a different matter and the Riley Scott was the starting point of at least three national designs of bombsight.

As noted earlier, Zeppelin airships carried passengers all over Germany in the final three years before World War I. In Florida a small flying boat began a regular service across Tampa Bay in 1914, taking one passenger on each trip; this was the world's first heavier-than-air airline. Daredevil American flyers, the first so-called barnstormers, thrilled thousands at shows in fields across the United States. In 1911 Calbraith Rodgers actually flew from coast to coast, but the epic 4,000-mile (6,437-km) trip took seven weeks, and the frail Wright flyer was repaired so often it was said the only original bits remaining at the finish were the rudder and the engine drip pan! Perhaps a greater accomplishment was that of Roland Garros, the pilot of the French Morane-Saulnier company, who on 23 September 1913 crossed the Mediterranean from St Raphael to Bizerta in 7 hours 53 minutes.

Not least of the accomplishments of pilots in 1911–14 was the mastery of a few basic forms of aerobatics. Prior to 1911 aero-

Right: Crashes, called spills, tumbles or upsets, were all in the day's work for early aviators. This Voisin was repairable. The tower in the rear is for launching Wrights.

planes could only just stagger into the air and had to be coaxed to maintain altitude. Apart from trying to climb or fly level under power the only other kind of flight was the glide, usually known by the French name *vol-plané*, to lose height prior to landing. As aircraft got stronger and more powerful their pilots dared to put the nose down more steeply and dive – though pulling out from a dive caused not a few crumpled wings, which were invariably fatal. Climbing too steeply caused the sudden loss of lift called a stall, and this in turn frequently resulted in the fatal condition called a spin. In a developed spin the aeroplane is actually stalled, so pulling back on the control column merely makes matters worse. The spin, which at first was mistakenly thought of as a spiral dive, was probably the number one killer of pilots in 1911–14. The answer was found by Lt Wilfred Parke of the Royal Navy who, in late 1912, got into a spin in an Avro 500 at the flying school at Chingford. He managed to recover and retraced his steps afterwards to discover what had happened. He had not tried to pull out of the dive but had stopped the rotation by using the rudder; then he had successfully pulled out of the ensuing straight dive.

A few bold pilots deliberately got into

spins and, following what Parke did, lived to tell the tale. Other truly courageous flyers deliberately pulled up into a vertical climb to make the so-called stall turn, where the machine almost stops in mid-air. Others made banked turns at 90° and progressed to make a complete 360° roll. But the manoeuvre which most caught the attention of the public was the loop, first performed either by the Russian Nesteroff over Kiev in early 1913 or by the French master of aerobatics Adolphe Pégoud. Certainly Pégoud was first to do the more frightening and dangerous outside loop, or bunt, which bends the wings downwards and tries to pull the pilot out of the cockpit.

Though not appreciated at the time, proficiency in aerobatics was soon to make the difference between life and death in air combat, although the official British view continued to discount the whole idea of war in the air and the trivial efforts of aeroplanes as bombers were held to reinforce this view. The most any British official or senior officer was prepared to admit in the summer of 1914, as the war clouds gathered, was that aeroplanes might have some small use in the reconnaissance role. There was little government response to public unease after the throbbing of German airships had been heard by night round the English coasts. On 5 August 1914 the British awoke to find they had declared war the previous evening, and the Royal Flying Corps' No 4 Squadron made its first combat patrol over Chatham with a flimsy BE.2 biplane which would have been unable to intercept a Zeppelin had one appeared.

Eight days later the first handful of RFC aircraft crossed the Channel, the first of more than 20,000 heading for France over the next four years. These early days were marked by what appear to us today to be ludicrous incidents. One pilot asked "What do I do if I meet a Zeppelin?", to be told "You should do your best to stop it". The only way this could be attempted was by a head-on collision. One whole squadron followed their leader down into a most dangerous ploughed field, not realising the CO had suffered an engine failure. Another British pilot was promptly arrested and locked up when he landed to ask the way, and this triggered off the idea of national markings. At first Union Jacks were sewn on the fabric of British aircraft, but later various target-type 'roundels' were devised for the Allied nations. The Central Powers used various forms of black cross with a white outline or background, Austria-Hungary additionally using bold red and white bands. In a sense, it was a reversion to the coat-armour of mediaeval knights, which could be read by every soldier in the army even though he could not read ordinary writing.

For a while there was a further similarity in that the opposing pilots used to wave to each other, seldom showing any animosity. From the earliest days of flying pilots had been a rather special fraternity, united in their skill, daring and exclusive 'shop talk' which bridged the usual problem of

Below: Pilot and observer of an RFC squadron at the front check the armoury. Roof beams carry Smith & Wesson revolvers and Very signal pistols.

language. This carried over into World War I, and while it was perfectly fair for aviators to toss small bombs on to ground targets and be shot at in return, it would initially have been considered unsporting to try to kill an enemy flyer.

In the first days of the war the almost total lack of experience of pilots on reconnaissance missions led them to report shadows of trees along French roads as columns of troops and groups of frightened civilians as fresh enemy battalions. This reinforced the general army view that aircraft were worthless, so that when vital reports came in from RFC aircraft in the Mons area in late August they were ignored. Not until it was almost too late was it realised the British Army was in a trap, and the famous 'retreat from Mons' was pulled off at the last moment.

German aviators were less in evidence because their armies were advancing rapidly on all fronts, but the appearance of their Taubes and Albatros machines proved too much of a temptation for the hot-heads of the RFC who succeeded in forcing down two of them on 25 August simply by aggressive flying. Lt Spratt did the same a few days later with his tiny, unarmed Sopwith Tabloid. Other German pilots were of sterner stuff, and on 30 August Lt von Hiddessen courageously flew his flimsy Taube monoplane all the way to central Paris, where he threw out some small bombs and leaflets before returning safely.

Aerial weapons at this time were still highly experimental and were often local lash-ups. The commonest were personal rifles and revolvers, which were often carried on front-line missions as morale-

Below: The aircooled drum-fed Lewis was used on most Allied two-seaters, aimed by the observer from a rotatable elevating mount.

Below: The point of this picture is the armament: four Le Prieur rockets are mounted on each of the V-type interplane struts of this Nieuport XI scout of the British Royal Naval Air Service. Above the upper wing is a mount for a Lewis machine gun, not fitted. The serial number N976 is not British but a Nieuport factory number; British serial was probably 3976.

976

boosters for the crew's protection. Bombs were by this time factory-made, though still small enough to be carried in or beside the cockpit and dropped by hand with no form of accurate aiming. Several aggressive pilots tried going aloft with machine guns, but the types readily available – such as the British Vickers – were so heavy that most machines could barely climb with them on board. The RFC had selected the much lighter Lewis gun, which also scored in being aircooled and using 97-round drums of ammunition, but few of these were available early in the war. The Germans and Russians used weighty machine guns similar to (and in the case of Russia, the same as) the Vickers, but the French Hotchkiss and Franco-American Benet-Mercier were lighter and fed by neat clips of ammunition. Flechettes were often used against troop concentrations or cavalry in 1914, until it was realized that these plain steel darts seldom hit anyone. A more sophisticated weapon was the Le Prieur rocket fired from the interplane struts against kite balloons and airships. The main problem here was learning how to aim correctly and judge the range, because the rocket followed an arching ballistic trajectory, rising to a peak and then plunging to earth.

Despite generally puny weapons, a few bold pilots were able to inflict damaging blows on the enemy. On 8 October 1914 there were just two British aeroplanes left in the beleaguered city of Antwerp, which was abandoned by the Allied armies. They were tiny 80-hp Sopwith Tabloids of the Royal Navy, one flown by Sqn Cdr Spenser Grey and the other by F Lt Marix. Each was loaded with four 20-lb (9-kg) bombs, and they set off in broad daylight for Germany. Spenser Grey dropped his bombs on Cologne railway station, while Marix flew low over the giant Zeppelin sheds at Düsseldorf and destroyed the airship Z.IX (which by coincidence had bombed Antwerp on 25 August, causing 26 civilian casualties). An even more audacious raid took place on 21 November 1914, when three Royal Navy Avro 504s dropped 11 bombs of 20-lb size on the main Zeppelin factory and gasworks at Friedrichshafen on Lake Constance.

Such raids then became rare on what deteriorated into a basically static Western Front, too far from Germany for such small low-powered bombers. But the number of aircraft on that front, and on those between Austria-Hungary and Italy in the south and Germany and Russia in the east, grew from dozens to hundreds and then to thousands. Most were engaged in reconnaissance, but on the vast Russian front the huge land armies occasionally saw the forerunners of a new species of aeroplane overhead; it was much larger than earlier types, with the capability of flying several hours and of carrying machine guns and heavy bombs. These were the great Sikorsky IM class bombers, so called because they were named for Russia's folk-hero Ilya Mourometz, which equipped the czar's EVK (*Eskadra Vozdushnykh Korablei* or 'squadron of flying ships').

Young Igor Sikorsky, much later to be

Right: A 'Bloody Paralyser' – in other words a Handley Page O/100 – is readied for a bombing mission on the Western Front.

Below: Zeppelin L9 was typical of the airships at the start of World War I. On 14 April 1915, in command of Heinrich Mathy, L9 bombed villages in Northumberland.

A 2

the chief inventor of the helicopter, built his giant four-engined *Le Grand* in 1912–13, and it was flown on 13 May 1913. Its success prompted him to design the first of the even bigger IM series, built at the RBVZ (Russo-Baltic Wagon Factory) in 1914 which was powered by four 100-hp Argus engines and had a crew of five. Not least of the surprising features were compartments in the rear fuselage for eating and sleeping, and a 'promenade deck' above the fuselage with a light rail round it on which crew-members could take a stroll while the machine was thousands of feet in the air! During the war about 80 of these great bombers were built; because of the chaotic supplies of materials, parts and engines there were many versions (it was common for each IM to have two pairs of engines of

Right: Although fire from the enemy trenches may not frighten the crew of this RFC BE.2c, they know that they would be easy target for any German fighting scout.

Above: Very few early-wartime aircraft had any provision for armament, but this RFC Bleriot XI-2 has a Lewis aimed (to the left) by the observer.

Left: C. H. Pixton reclines against the tiny Sopwith Tabloid seaplane at Monaco in 1914 immediately before winning the second Schneider Trophy race for Britain.

different types!). They opened their strategic bombing campaign on German targets on 15 February 1915 and made some 400 raids. Only one was shot down but only after its gunners had shot down three German fighters. IM bombers carried about 1,500 lb (680 kg) of bombs on missions lasting five or six hours, and has as many as seven defensive machine guns.

The other country to pioneer the heavy bomber was Italy, where Count Gianni Caproni was logical enough to realise that since a few tiny bombs dropped on the Turks in 1911 had had a large effect, it would be sensible to design a much bigger aircraft specifically as a bomb carrier. Like Sikorsky, he flew his first large biplane in 1913, but adopted an odd engine layout, with three Gnome rotary engines in a short central nacelle connected to a pusher propeller at the rear and two tractor (pulling) propellers on the wings. In October 1914 a second machine flew with a better layout, with a central pusher engine and two tractor engines on the wings, coupled directly to their propellers. This Ca 1 was a great success and 162 were built by 1916, forming the world's most powerful striking force in the first two years of the war. Later, 270 of an improved model were built, plus 86 built in France for the French Aviation Militaire. These tough and capable Capronis made their first big bombing raid against the empire of Austro-Hungary on 20 August 1915 and were the greatest pioneers of heavy raids on strategic targets. It is difficult to express in words the problems of flying for up to seven hours in the freezing cold across such dangerous terrain as the Alps, the gunners wrapped in leather coats as they stood in the open within inches of the giant propellers.

In Britain it was the RNAS (Royal Naval Air Service) which took the lead in heavy bombers. In December 1914 Cdr Murray Sueter, the dynamic Director of the Admiralty Air Department, asked designer Frederick Handley Page to build 'a bloody paralyser of an aeroplane' to hit Germany and ships far out at sea. By this time the Royal Ordnance Factories were producing a substantial bomb weighing 112 lb (51 kg), and Sueter asked Handley Page to make a bomber able to carry six of these. The Admiralty had already asked the Rolls-Royce car company to produce a really good and powerful aero engine because there were none in Britain. Almost all the Allied engines were French, and many of these were not very good. After considering using other types of engine, Handley Page got the rough specification of the new engine, designed to give 200 hp, from Henry Royce and set about building a bomber with two of them. At last the first O/100 bomber was completed at the Cricklewood factory in London. It was wheeled out secretly on to the Edgware Road and towed by night to the aerodrome at Hendon where it flew on the following day, 18 December 1915. Instead of six of the big bombs, it proved able to carry 16! Subsequently 40 of these great machines were made, plus six with 300-hp Sunbeam engines; they were followed by no fewer than 550 of the improved O/400 type, most of which had later versions of the Rolls-Royce engine, the Eagle VIII, rated at an impressive 360 hp.

The great Handley Pages made increasingly damaging raids on German targets, but it was only in the final three months of the war that they operated in large numbers. From September 1918 the heavy bomber squadrons of the recently formed Royal Air Force received a 'block buster' bomb weighing 1,650 lb (748 kg), and these devastated several cities in western Germany. Meanwhile, Handley Page had built an even bigger bomber, the V/1500, to equip a new Independent Force to carry out heavy attacks on targets as far away as Berlin. The monster V/1500 flew in May 1918 and, while squadrons were formed to use it, a giant bomb of 3,300 lb (1,497 kg) was put into production, but the Armistice came on 11 November 1918 before the first raid took place. The main operational mission of the V/1500 was a single very difficult raid over the Himalayas in May 1919 to wild Kabul, the government of which was so amazed at the bombing of the royal palace that it immediately brought the Afghan war to an end!

With aircraft able to inflict such blows it was obvious not only that anti-aircraft guns would improve but also that fighter aircraft would be developed to shoot down enemy machines in flight. In 1914 this was not easy to do. Pilots thought of all kinds of desperate measures and tried them out.

Right: In 1914 French gunners were told to shoot at anything with an engine in the nose. This Caudron G.III, a type used in vast numbers by all Allies on the Western Front, would have probably been safe!

Right below: Another pusher was the British Vickers F.B.5, known as the 'Gun Bus'. The observer in the nose had a Lewis, not fitted to this fine modern replica.

Below: Believed to be serving with Escadrille 1 of the Belgian Aviation Militaire in 1914, this Farman HF.20 has a Lewis gun in the nose.

Below: One of the best fighting scouts of 1916, the Nieuport XVII, continued the fashion for a small lower wing. It was so agile the Germans copied the idea in their Albatros series.

Right: The ubiquitous Lewis machine gun was usually stripped of its tubular air-cooling cover when used from aircraft.

One threw a brick at an enemy pilot but missed. Capt Kazakov, of the Imperial Russian 19th Squadron, became famous for trailing a hooked grapnel on a long wire. Flechetts have already been mentioned and were tried along with small bombs against both enemy aeroplanes and airships, but the general feeling continued to be that the most effective weapon was the machine gun.

The difficulty was how to mount it. Apart from the fact that the weight of such a gun and its ammunition was almost too much for 1914 aircraft, the normal tractor aeroplane with the propeller in front precluded a gun firing straight ahead, because it would shoot off its own propeller. The alternatives were either to have a second man in the aircraft to shoot the machine gun towards the rear or to design the aircraft as a pusher. In fact, so many of the Allied aircraft in the first year of the war were pushers (especially French) that it was common for anything with a tractor propeller to be fired on as being an enemy! This is despite the fact that Britain built the first real fighter, a pusher, as early as 1912. It was called the Vickers EFB (experimental

Right below: The standard German gun was the belt-fed 7.92-mm Spandau, here seen fitted to one of the 1915 Fokker E-type monoplanes.

Built by Oakley Ltd in Ilford in 1916, this Sopwith Triplane is the genuine article, not a replica. There was no magic about using three wings, but by April 1917 the RNAS had used their handful to such effect that, according to the official historian, 'the sight of a Triplane formation caused the enemy to dive out of range'.

fighting biplane), and had a short central nacelle with a pilot and a Maxim (Vickers) machine gun in the nose, aimed by an observer. Boldly, Vickers began building 50 of an improved version in 1914, even though they had not been ordered, and these and many others later achieved fame as the Vickers 'Gun Bus'.

Early in the war there were many pusher aircraft armed with a nose machine gun, and one particular machine was Voisin V.89 of the French Aviation Militaire. The Voisins were frail-looking pushers, seldom able to reach a speed as high as 70 mph (112 km/h) and seemingly useless. Despite their four perambulator-like wheels and quaint appearance they were made of steel and were so tough they stayed flying all through the war mainly on bombing, reconnaissance and training missions. But on 5 October 1914 V.89 set off with Sgt Joseph Frantz piloting and Cpl Louis Quénault in the front manning his Hotchkiss machine gun. They chanced upon a German reconnaissance machine, thought to have been an Aviatik, and managed to get near enough to it for Quénault to open up with his gun. Suddenly the 'Hun' pitched down out of control and was seen to crash, the first aeroplane ever shot down by another.

Below: Although many German pilots of World War II scored over three times as many victories, they will never displace the great Richthofen (Red Baron) as the most famous ace of all time.

Occasional 'victories' followed, but they were rare and hard to achieve, so there was a great sensation when Garros, the famed pre-war Morane-Saulnier pilot, shot down five Germans in the first 18 days of April 1915! In 1912 Raymond Saulnier had experimented with an interrupter mechanism to enable a machine gun to fire ahead past the blades of a tractor propeller, but there was no way to keep the engine at exactly the speed needed, and shooting off one's own prop blade could be lethal because the imbalance could tear the engine out of the aircraft. There was no official interest and the idea was dropped. Garros and Saulnier instead fitted heavy metal triangular deflectors to the blades of a Morane, so that it did not matter whether the blade was struck or not. With this crude scheme Garros was able to shoot down enemies by aiming the whole of his tractor aircraft, with the results that made headlines.

Unfortunately for the Allies he was himself brought down on 19 April and the indestructable deflectors revealed all. The top designer for the Germans, Dutchman Anthony Fokker, was asked to copy the idea, but instead he quickly came up with a proper interrupter gear that worked reliably. He demonstrated it on one of his agile little E (for *Eindecker*) monoplane aircraft, low-powered machines that had been flying in prototype form in 1913. The German staff insisted that Fokker himself should don a uniform and try the idea out on the Western front (though he was technically a neutral person). Whilst there he made the acquaintance of Oswald Boelcke, who tried out the Fokker with the forward-firing gun. He quickly began scoring like Garros, and became Germany's first ace (originally, a pilot with five air victories). A brilliant leader, he worked out the entire basis of aerial combat, trained junior pilots and put the quite flimsy low-powered monoplanes in a position of such ascendancy that British pilots were angrily described in Westminster as mere 'Fokker fodder'.

Britain and France had to hit back. Already there were the little French Nieuports with a Lewis gun firing from the top wing above the propeller, and various British pushers such as the neat DH.2 and the big but tough two-seat FE.2 series. But it was not until 1916 that really good fighting scouts began to appear on both sides. Among the best were the British Sopwith Pup and Camel, SE.5a and the two-seat Bristol Fighter, the French Spad family and the German Albatros biplanes. The Sopwith company built a few triplanes,

Left: The Albatros C.III two-seater was a tough but slow machine used for reconnaissance and light bombing throughout the Western Front.

with three wings, for the RNAS which used them brilliantly. They created such a stir that many German and Austrian firms copied them, and though only a few were made the Fokker triplane, the Dr.I, is famous because it was one of the favourite mounts of the greatest ace of all, Baron Manfred von Richthofen.

Popularly called The Red Baron today because of the colour of his triplane, von Richthofen was a great leader who formed his pilots into enormous loosely arranged

Below: The unrefined cockpit of an early Sopwith, where wires created more problems than the lack of instrumentation.

groups called Circuses. He allowed each pilot to paint his machine in bright individual colours and when the whole swarm dived to attack it was a stirring sight. In contrast the RFC aircraft were painted a deep khaki all over, and because they flew in squadrons of nine to 12 they were often greatly outnumbered. But by skill and courage they held their own and by 1918 – despite the appearance of the Fokker D.VII, the best of all German fighters – the Allies had air superiority.

Every flight was a tough test of man and machine. Although more powerful engines meant that there was no longer much doubt about becoming airborne, failures of the engine and fuel supply were still common and there was no way most pilots could fly in bad weather. Even night flying was in its infancy and some pilots crashed when friendly searchlights were aimed at them, blinding them as they were landing, instead of illuminating the field ahead. Most combat missions involved a prolonged climb to 10,000 to 20,000 ft (3 to 6 km) altitude. There was no cockpit heating, no heated clothing, no oxygen and no parachute – and usually no armour either. It needed plenty of guts to go out to meet the machine guns of the enemy.

This Fokker Dr.I triplane may be the scarlet machine in which 'Red Baron' Manfred von Richthofen was killed on 21 April 1918.

In 1921 a young engineer named Donald W. Douglas set up shop as a planemaker and managed to sell this chunky single-seater to the Navy, which called it the DT-1. It was good, and led to many more sales – and also to a special version for the Army whose objective was to make the first flight round the world (which was achieved).

4. THE TRAIL-BLAZERS

THE 1920s OPENED with large numbers of ex-wartime pilots, grateful still to be alive but out of work. Europe, the USA and some other regions were littered with tens of thousands of unwanted aircraft, many of them unused. In many countries some of the surplus men and machines came together and struggled to make a living. In the USA they formed a new generation of barnstormers, far more numerous than the old and generally even more dardevil than before (because fresh aeroplanes cost next to nothing – for a while). Deliberately crashing an aeroplane became a popular stunt for airshows and for the fast-growing business of making movies in sunny Hollywood – where, because of the fine weather, aircraft manufacturers began to gravitate, among them Donald Douglas, John K. Northrop and the Loughhead brothers, who soon spelt their name as it was pronounced: Lockheed.

In Britain former majors and colonels found it hard to settle down in an office, and were glad to earn a few shillings back in the cockpit, hopping from fields near towns, or from beaches and anywhere else they might find customers for pleasure flights. More serious was the struggle of the infant airlines. Regular passenger services had begun while the war was still on in Hungary and Germany, and carrying the mail between New York and Washington. The RAF opened services between London and the Paris Peace Conference, and to Cologne, while even in the chaos of civil war in defeated Germany several commercial lines were operating in 1919. And on 25 August 1919 ex-Major Cyril Patteson took off in his DH.16 from the grass at Hounslow – the aerodrome for London, next door to today's Heathrow – to fly to Paris on the first flight of the world's first sustained international daily scheduled service.

The DH.16 was a biplane powered by a single 360-hp Eagle engine, the same as powered the big Handley Page O/400 bomber. The Handley Pages were also converted to carry passengers instead of bombs. The DH.16 was a conversion of the DH.9A light bomber, completely redesigned to carry four passengers in a cabin behind the pilot's open cockpit. This was twice as many as the two-passenger DH.4A, but it was possible to buy a 4A for about £1,200, whereas a newly built 16 cost £4,000. So the airlines started off wondering whether to buy cheap, converted warplanes or expensive civil airliners.

Very soon, although the network of air routes grew across Europe, and began to make embryonic appearances elsewhere, the airlines found the going so hard that many went bankrupt. Governments had previously proclaimed how important civil aviation was going to be, but when it came to the unpopular task of spending money they soon lost interest, except for a few who saw that subsidizing their own airlines might be a good investment. The basic problem was that the primitive airliners were generally unable to operate at a profit at a level of fares low enough to attract customers. They carried too few people, flew slowly (always well below 100 mph, 161 km/h), and generally were uncomfortable.

Many of the early civil machines had open cockpits for the passengers as well as the pilot. Customers were advised to dress in their heaviest coat and warmest clothes,

Below: Wing-walking on a Curtiss JN-4 was just one of the acts in the early barnstormer's repertoire. If the crowd was big enough he might even stage a crash.

Left: One can almost hear the steady humming drone as US Navy airship ZR-3 USS Los Angeles, *cruises past the tip of Manhattan around 1930.*

wear heavy gloves and if possible a leather helmet and goggles! Air travel was a very daring experience for the 1920s equivalent of today's Jet Set. Europe, where nearly all the struggling little airlines were based, was already covered by a network of railways which were cheaper than the airlines and much more reliable. And there was not much time difference on most routes between a train at 60 mph (96 km/h) and an airliner at 80 mph (129 km/h). To try to attract more traffic, some airlines reduced fares; but unless they enjoyed a government subsidy they simply lost even more money. Paring costs to the bone, they cut down on wages and skimped on overhauls and maintenance, which did not help the basically unreliable wartime engines. In each of the first five years of the 1920s there

Below: Demonstrating the Vimy Ambulance, derived from the famous bomber which made great flights in 1919–20.

Above: In turn the RAF Vernon led to the Victoria, an 80-mph (128.8-km/h) troop carrier used mainly in the Middle East throughout the inter-war period.

were far more airline forced-landings and crashes in Europe than in the five years 1979–83, when traffic was more than 1,200 times greater! Two passenger airliners even collided head-on between London and Paris, because they were both following the same railway line, and at that time there were no rules about flying at different heights depending on the direction of travel.

Initially France led the way in civil aviation, and thanks to strong government support, a network of powerful companies opened services with landplanes and flying boats (seaplanes with a boat-type hull) across Europe and over the Mediterranean to North Africa. It seems amazing today to realise that, should an aircraft be forced down in French North Africa in the early 1920s, those on board might well be tortured to death! Despite this, the routes in Africa grew and gradually spread south, and what was generally just called *La Ligne* (the Line) reached Dakar in June 1925. On 1 March 1928 the great French dream of reaching South America came true, with the opening of the through route to Buenos Aires (but with Dakar-Natal linked by ship). In 1929 the route battled on over the towering Andes to Chile, and in May 1930 Jean Mermoz at last flew a Latécoère 28 seaplane on the Dakar-Natal sector. Regular air service right through from Paris to Buenos Aires began in May 1934.

Towards the east the French aimed at distant Indo-China, finally completing the route to Saigon in January 1931. Here there was no long ocean crossing, and other nations helped to establish the longest air routes of the 1920s. The Dutch airline KLM was founded in October 1919 (today it boasts one of the longest periods of unbroken existence of any airline), and thanks to the establishment in Amsterdam of the post-war Fokker factory was able to use excellent purpose-designed transports from the outset. As early as 1924 a Fokker F.VII managed to fly all the way to Batavia, biggest city of the Dutch East Indies, though it was to be another seven years before a regular service opened.

Britain was most ambitious in its long-term plans; the vast Empire cried out for speedy links using what Cayley a century earlier had called "The great ocean of the air, that comes to every man's door". But the planning was easier than the execution. In the immediate post-war years it was a great achievement just to have flown an intercontinental distance, even if it took weeks.

Britain certainly led in such pioneering flights, although the first aeroplane flight across the Atlantic was made by a giant US Navy flying boat, the Curtiss NC-4, commanded by Lt-Cdr A. C. Read, which crossed in stages to Lisbon and then Plymouth between 8 and 31 May 1919. Apart from this, the pioneer long-distance flights were by British or Commonwealth pilots. The most famous of all was the first direct (non-stop) crossing of the North Atlantic. Many pilots tried to accomplish this in 1919 and some vanished in the grey wastes. However, on 14 June 1919 a Vickers Vimy, one of the latest RAF bombers powered by the trusty 360-hp Rolls-Royce Eagle engine, set out from St John's, Newfoundland. It was crewed by Capt John (later Sir John) Alcock and Lt Arthur Whitten Brown who, despite hair-raising difficulties, maintained the remarkable speed of 115 mph (185 km/h) over the 1,890-mile (3042-km) route, landing next day at Clifden, Ireland.

Only a few days later the RAF airship R.34 slipped its moorings at East Fortune, Scotland, and flew non-stop to Mineola, New York, where it arrived on 6 June. Three days later she cast off again and, assisted by tail winds, arrived at Pulham, Norfolk, five days later, having completed the first double crossing of the Atlantic. This was the first such flight that did not appear a foolhardy adventure. Even after this date, several more aviators went to a watery grave in the Atlantic, which continued to pose a major challenge to aeroplanes until after World War II.

A Vickers Vimy also made the first flight from Britain to Australia. The Australian brothers Capt Ross Smith and Lt Keith Smith took off from Hounslow on 12 November 1919, and pioneered the route that so many others were to follow later, arriving at Darwin on 10 December 1919. Of course, this was not the first aeroplane in Australia, but it was the first to fly there. Two days after the Smiths landed they were joined by Capt Wrigley and Lt Murphy who, in a flimsy little BE.2e, had been the first to fly across Australia, having taken 46 hours flying time to travel from Melbourne, which they had left on 16 November.

Even more protracted was the journey of Lt-Col Pierre van Ryneveld and Sqn Ldr Christopher Brand, who took off in a Vimy from Brooklands, England, on 4 February 1920 to fly to their native South Africa. On 6 March they crashed at Bulawayo, Rhodesia, and finally got hold of a much smaller DH.9 in which they arrived at Wynberg aerodrome, Cape Town, two weeks later. The routes to Cape Town and to Australia were among the longest in the world, yet the British government never lost sight of a scheme of great Imperial air services to knit the far-flung Empire together.

In the early 1920s, since aeroplanes seemed relatively puny, and totally unable to maintain air links with Canada, the future seemed to lie with airships. There was a major hiccup after 24 August 1921, when the newly-built British airship R.38 broke up in the air and exploded, only five of the 49 people on board surviving in the

Below and right: In 1919 designers were busy turning bombers into airliners. Few were bigger than the Italian Caproni Ca.48 triplane with 23 (later 30) seats.

AUVER

Above and left: The Farman Goliath's wings were 'made by the mile, and sawn off as needed'. They gained an enviable reputation with French airlines from March 1920.

tail section. The R.38 was based on the latest wartime Zeppelins, and the Zeppelin company was allowed to continue its work in Germany, but the giant new airships it built were handed over to the Allies as war reparations. Biggest of these was the LZ.126, which flew non-stop to the USA to become the US Navy's *Los Angeles*. She made 331 flights before being retired in 1932. One reason for her safe flying may have been that she was designed by an experienced German team and filled with helium. This non-inflammable gas was almost a US monopoly. The US Navy commissioned two gigantic airships in the 1930s, the *Akron* and the *Macon*, but both had quite short lives and crashed into the ocean.

Other nations also had trouble with airships. France's *Dixmude* vanished with all hands somewhere near Sicily in December 1923. Italy's *Italia* crashed near the North Pole on 25 May 1928, flinging out most of her crew onto the jagged ice; she then rose into the sky, with six men still aboard, never to be seen again. Even Britain continued to be reminded of airship problems; on 16 May 1925 the R.33 was torn from her mooring by a gale and driven as far as Holland before the skeleton crew aboard managed to bring her back.

Despite all this, Britain decided to build two giant civil airships to open the planned scheme of Imperial communications. After two years of talking, plans at last went ahead in 1924. One ship, the R.100, would be built by private enterprise, in Yorkshire by Vickers at Howden. Its rival, the R.101, would be a state undertaking by the Air Ministry at Cardington, Bedford. They were to be ready in 1927. In fact both ships were two years late, and to the government's embarrassment the R.100 proved to be better. She maintained a speed of 81 mph (130 km/h), the fastest airship in history, and made a fine round trip to Canada in the summer of 1930. Her chief designer was B. N. (later Sir Barnes) Wallis, one of the few modern British engineers whose name became a household word, and the chief stressman was N. S. Norway who also became famous as the novelist Neville Shute.

Tragically, the R.101 – though in most ways a more advanced design – was riddled with faults, many of which were hastily bodged because the government thought time more important than safety. The Air Minister insisted she was ready to fly him to India and back in September 1930, so that he could attend a big Imperial Conference in London on 20 October. All the experts said the ship was not ready, but the government brushed aside such trivialities, and R.101 set off from Cardington at dusk on 4 October with a temporary Certificate of Airworthiness on board which stated that the airship's vital tests would be done during the journey to India! Losing gas, sodden with rain and suffering from other faults, she slid quite gently into a hillside near Beauvais, France, and burst into flames from stem to stern. Only six of her 54 occupants escaped, not including the Minister. Her rival, which was genuinely airworthy, was immediately broken up and sold as scrap for £450.

This saga of government mismanagement really set the tone of the relationship between the British government and its aircraft industry for the next half-century. In Germany, however, the Zeppelin company was at last permitted to build ships larger than the 1,100,000 cubic feet (30,800 m³) limit imposed by the Allies, and on 8 July 1928 the LZ.127 was christened *Graf Zeppelin* in honour of the long-dead pioneer. She was destined to make over 650 great voyages, carrying more than 18,000 passengers on memorable journeys to all parts of the world, but mainly to South America. Sadly, the explosion of her big sister *Hindenburg*,

Below: Alcock and Brown (right) piloted this Vimy on the first non-stop Atlantic crossing in June 1919. Too late they realised they were landing in a soft peat bog!

described in the next chapter, made the Germans decree she could no longer fly with hydrogen; and because the Americans refused to export helium, *Graf Zeppelin* was scrapped in 1938. She was the only profitable rigid airship ever built.

After the pioneer long flights of the 1919–20 period, few flyers made the headlines again, except for the competitors in the Schneider Trophy race. Jacques Schneider, a noted arms manufacturer, donated the giant trophy in 1912 to encourage speed and seaworthiness in seaplanes. The second factor was quickly forgotten and the annual race – interrupted by World War I, and abandoned in 1919 because of fog – became the number one international aviation race. In the early 1920s lumbering flying boats were winning at speeds below 150 mph (241 km/h), but in the United States Glenn Curtiss, who had taken the main speed prize at Rheims in 1909, opened the way to racing machines faster than any previously built.

Much of the credit was due to his engine assistant Charles Kirkham, who designed the D-12 engine. This was a Vee-12 (12 cylinders in two blocks arranged in V shape seen from the end), and it pointed the way ahead by having each block of six water-cooled cylinders cast from one piece of aluminium. Not only was it light and powerful but Curtiss streamlined its installation to reduce drag, so in the period 1922–25 Curtiss landplane and seaplane racers set many speed records and twice won the Schneider Trophy. Under the rules, any nation winning three times in a

Main picture: First flown in July 1933 France's Dewoitine D.332 looked the last word in airliners, and is seen on its first visit to Croydon.

row could keep the trophy in perpetuity, but in 1924 the American hosts sportingly declared the contest off when opposing teams said they were not ready. Famed pilot Jimmy Doolittle, who in 1922 had made the first US coast-to-coast flight in one day, won the 1925 event at 233 mph (375 km/h), but in 1926 a massive effort by Italy with one of the new monoplane Macchis with a Fiat engine ensured that the 1927 race would be in the holiday atmosphere of the Venice Lido. Unexpectedly, this year it was Britain's turn, with the Supermarine S.5 designed by Reginald Mitchell, and powered by a boosted Napier Lion.

The race was changed to being held on alternate years and the 1929 race was a walkover by Britain, who did not follow the lead of the Americans in waiting for the opposition to get ready. This time the winner flew an S.6, bigger than the S.5 and powered by the specially created Rolls-Royce R engine, the most powerful ever made at about 1,900 hp. The 1931 race nearly did not happen at all. Because of the recession, Britain's economy-minded government refused to sponsor the 1931 race, and it was left to wealthy Lady Houston to write a cheque for £100,000 to allow Britain to compete! For the 1931 race Macchi was working on a fantastic new seaplane with two Fiat engines in tandem, but it was not ready in time and for the second time the British eschewed sportsmanship in favour of winning, and this time it was for keeps. The S.6B, with an R engine boosted to an amazing 2,500 hp, also set a world speed record at 407.5 mph (655.8 km/h).

These unprecedented speeds were certainly not reflected in the performance of ordinary aircraft. Fighters began the 1920s at speeds around 150 mph (241 km/h) and progressed to about 185 mph (300 km/h) by 1930. Commercial transports began at about 90 mph (145 km/h), and by the end

Left: The first big design job handled by Barnes Wallis was the giant 'private enterprise' R.100 airship, built by Vickers at Howden, Yorkshire, in 1929.

Right: The Vickers-built R.100 at her mooring mast at St Hubert, Montreal, during her very successful trip to Canada in July/August 1930.

Right, below: The stricken Hindenburg, *filled with hydrogen, was consumed in seconds as she was approaching the Lakehurst mast on 6 May 1937.*

Below: One of R.100's three power cars, each with a push/pull pair of 650-hp Rolls-Royce Condor engines. These burned more fuel than R.101's diesels, but were lighter.

of the decade were still seldom doing more than 115 mph (185 km/h), although there were some notable exceptions.

Throughout the decade there were literally hundreds of aircraft manufacturing companies – there were more than 100 in France alone – but most struggled for existence and only a handful had anything that could be termed a production line.

Left: It seemed natural to make the accommodation in an airship resemble a ship. This was the Grand Salon of the R.100, made of duralumin and mahogany.

Below: The Graf Zeppelin, *funded by a public subscription in 1926, carried passengers and cargo from 1928 until World War II.*

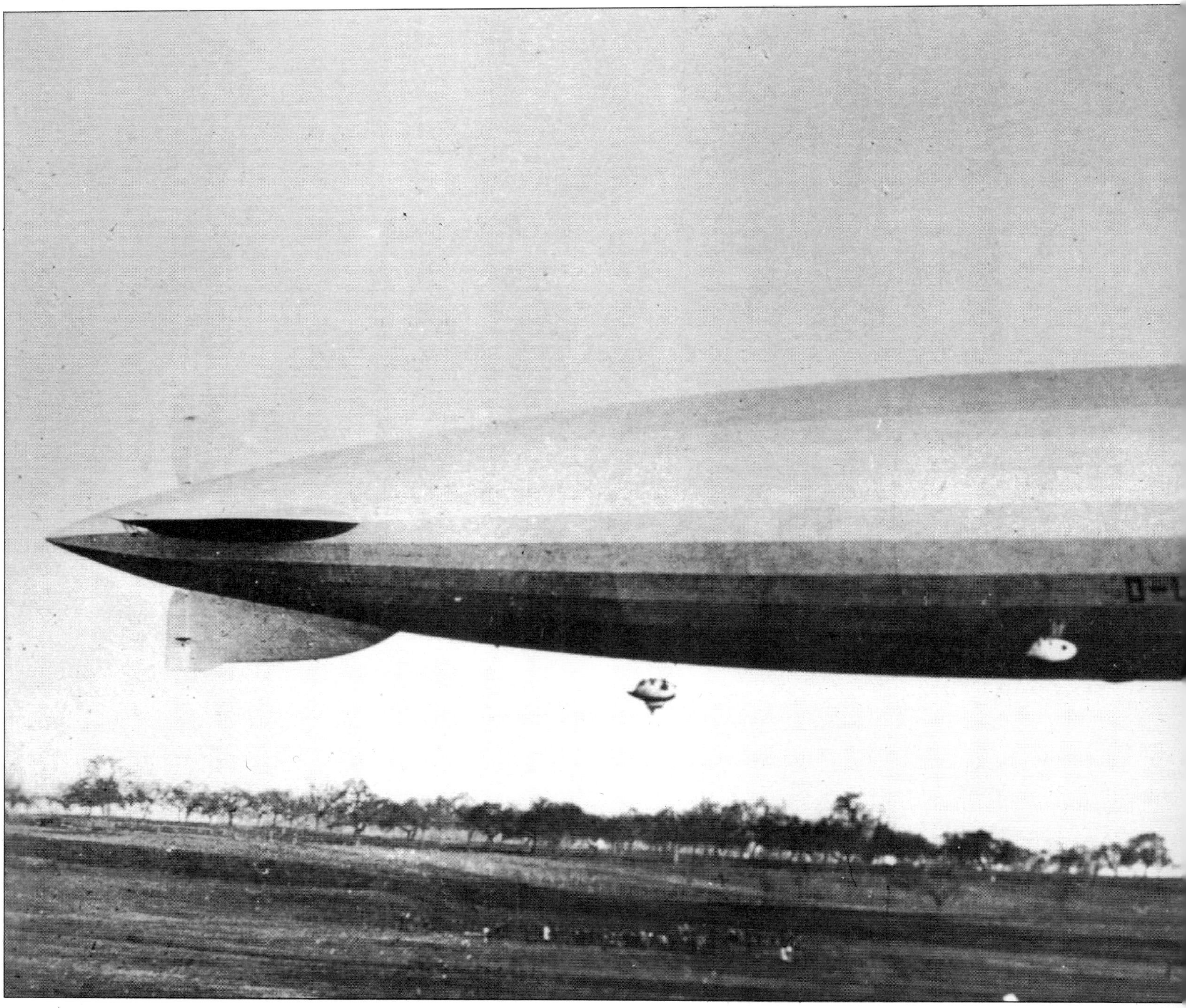

Left: This Supermarine S.6B was the aircraft which won the 1931 Schneider Trophy race and later set a new world absolute speed record.

Where transports were concerned, two European companies were undisputed leaders, with sales measured in hundreds. The first was Junkers in defeated Germany; the second firm was that of Anthony Fokker, who in 1919 had brazenly shipped trains filled with aircraft and parts from his wartime factory at Schwerin, north of Berlin, to his new site at Amsterdam under the noses of Allied inspectors who never asked for the trucks to be opened.

Fokker was boastful but also highly competitive, and throughout the 1920s he sustained a massive output of fighters and

Above: Ford Tri-Motors were made in many versions, but all set a new standard of long-lived toughness. The interior (right) was typical of 1925–33.

light bombers which were sold all over the world. But his airliners were the ones that gained fame; they all had the same basic construction. The angular box-like fuselage was made of welded steel tubes, covered in fabric. Above it was mounted a thick wooden wing, with a load-bearing plywood skin, which needed none of the mass of struts and wires seen on other monoplanes. His F.II single-engine four-passenger machine flew in 1919, and via the five-passenger F.III came the six-passenger F.VII of 1923 with a wartime 360-hp Eagle engine. Such engines were naturally much cheaper than new designs, but it was gradually realised that more modern engines were designed not for war but for peace, where reliability over many years was a prime consideration. Nobody knew this better than Roy Fedden of Bristol, who in the final year of the war had run a new engine called the Jupiter.

Unlike most engines of the day, the Jupiter was an aircooled radial, with its nine cylinders arranged like spokes of a wheel. It was neat and relatively light, but thanks to Fedden's colossal drive and leadership at the newly formed Engine Department of the Bristol Aeroplane Company the Jupiter steadily became the number one engine of the 1920s. By the end of that period the Jupiter was being manufactured under licence in 17 foreign countries, and it powered 262 different types of aircraft. Thanks to its quality and reliability it achieved this position of dominance despite severe competition from the Pratt & Whitney company, whose first Wasp engine was run at Christmas 1925. Rather smaller than the Jupiter, the Wasp was again a nine-cylinder radial and the start of what was to become the world's biggest aero-engine company.

Both the Jupiter and Wasp were to be important in powering the high-wing Fokkers. The F.VII with the Jupiter was designated F.VIIA, and it flew in 1925 with two additional passenger seats, a total of eight. But by this time the safety advantages of multiple engines were obvious and later in 1925 Fokker flew the first F.VIIA/3m, with three engines. Naturally each engine had to be of lower power, and instead of the 400 or 450-hp Jupiter the 3m version had three Wright Whirlwinds of 200-hp each. Imported into the USA, the 3m won a special reliability trial organised by the Ford Motor Co, which was looking for a new airliner to carry mails and passengers on new routes opened up by the decision of the US Postmaster General to let private firms bid for mail routes.

Previously, from 1918, the Post Office itself had carried the mail. There was not a tougher job in history than that routinely undertaken by the early US Mail pilots. Flying slow, open-cockpit wartime bombers, such as the DH.4M (a US version of the British de Havilland design), they would load up with heavy sacks of mail in all kinds of weather, day and night. Eventually the network of trunk routes crossed the Rockies and reached the West Coast, and the flying conditions in winter were unbelievably severe. Bonfires lit on the ground at night provided some kind of guidance

Below: Britain's winning team at the 1927 Schneider Trophy race, Venice, with F/Lt S. N. Webster and S.5 N220. Reg Mitchell (centre) later designed the Spitfire.

until, towards the end of 1924, the main route between Chicago and Cheyenne was equipped with 30 emergency landing grounds with boundary markers, rotating light beacons and a telephone, while the route itself was marked by 292 flashing beacons. The aircraft themselves began to be equipped with luminous instruments, navigation lights on the wingtips and tail, a powerful pilot-switched landing light, and parachute flares for illuminating the ground when looking for a place for an emergency landing. Towards the end of the 1920s the US Mail routes had become the first members of today's global system of Airways, with radio beacons at every intersection and at intervals along each route so that regular services could be maintained in all weathers. The radio signals were sent out along the routes themselves, so that airline flying became known as 'flying the beam', the beam being heard but not seen.

Ford was one of the chief operators of early mail routes, and it quickly got into the business of building its own airliners. Ford bought the business of William B. Stout, which built all-metal monoplanes. Stout added two more engines to create the Ford Tri-Motor, first flown in June 1926. Noted for its corrugated metal skin, the Tri-Motors were built in large numbers, and were popularly called the Tin Goose. They were so tough that many survived more than 30 years of intensive use and a few are still flying today. Meanwhile, the corresponding tri-motor Fokker was given a great publicity boost by an Australian pilot, Charles (later Sir Charles) Kingsford Smith. He made many of the greatest pioneering flights, one being the first crossing of the Pacific, between 31 May and 9 June 1928. He and a crew of three flew his F.VIIB/3m *Southern Cross* from Oakland, California, to Brisbane (where this famous machine is still on display).

Fokker set up a second production line in the USA, which gradually gained its own design capability. In 1929 the US Fokker team flew the F.32, the first American four-engined transport with seats for no fewer than 30 passengers. The engines were arranged in push/pull tandem pairs on struts under the wing in what was then a common arrangement for very large aircraft. Ten F.32s were built, but the general opinion was that a 30-seater was much too large.

In terms of numbers sold the Junkers company of Dessau, Germany, was the top builder of civil airliners in the period 1920–35. This was because of the pioneering structural enterprise of Dr Hugo Junkers in using an all-metal airframe, with a skin of corrugated sheet. Such sheet could be made thinner than flat sheet without buckling, but it was not until much later that it was realised that instead of flowing neatly along the corrugations, which were always arranged fore-and-aft, there were places where the air actually flowed diagonally across them, and thus the corrugations caused significantly increased drag. Even if the customers had known this, it was not very serious at speeds under 100 mph (161 km/h), and the robust toughness of all Junkers aircraft was far more important.

A few early Junkers types served as close-support bombers in 1917–18, but the first civil transport was the F.13 of June 1919. In every way except its wartime water-cooled engine this was a truly modern machine. Not only was it made entirely of metal but the low-mounted monoplane wing was completely free of any external bracing, and the fuselage with seats for a pilot and four passengers was quite well streamlined. With an obsolescent BMW six-cylinder engine of 240 hp it was rather slow, at about 75 mph (120 km/h), but the F.13 was still the best small airliner in the world in the early 1920s and no fewer than 350 were successfully sold in all parts of the world.

Junkers continued development of the single-engined machines with the F.24, W.33 and W.34, powered by engines of some 400 hp such as the Jupiter, or Junkers' own in-line engines. But Junkers was one of the first designers to recognise that for commercial use it was very important to be able to continue flying (even if only just) after the failure of one of the unreliable engines of those days. At that time a twin-engined machine able to keep flying on one engine would have been hopelessly uneconomic; the airlines could barely

Below: This US-built DH-4, with Liberty engine, typifies aircraft which carried the US Mail during the pioneer years to 1927, though with a modified single-seat cockpit.

make money with their existing machines. But there did seem to be a case for using three engines, one on the nose and the others on the monoplane wings. Each engine then only needed to be a little more powerful than strictly necessary, for flight to be possible with one stopped. The result was the Junkers G.23, first flown in late 1924. To beat the Allied restrictions on Germany, Junkers had opened a subsidiary in Sweden, AB Flygindustri, and the G.23 was built here as well as in Germany later on. It went into service with the main Swedish airline on 15 May 1925, and though it had only eight passenger seats it sold well and led to the G.24 ten-seater of 1925 which had more powerful engines. From this stemmed the Jupiter-engined G.31 of 1926 and the Ju 52/3m described in the next chapter; but first, in 1929, Junkers built the remarkable G.38.

Back in 1910 Junkers had sketched a giant airliner that was almost all-wing. In the late 1920s he finally built it, and the G.38 was the result. Not only was it extremely large and of all-metal construction, but six of its 34 passengers occupied cabins in the leading edge of the thick wing, inboard of the inner engines, which was skinned with celluloid and glass!

Very large aircraft were rare in the 1920s. Britain built the Beardmore Inflexible in 1928, with three 650-hp Rolls-Royce engines and a span of 157 ft 6 in (48 m), to test a system of construction, but it achieved little. In 1929 Caproni in Italy took his inverted-sesquiplane (a biplane with the upper wing much smaller than the lower, the reverse of usual practice) bombers to the limit with the Ca 90. This had about the same span as the Inflexible but a much greater gross weight of 66,140 lb (30,000 kg), and was powered by six 1,000-hp engines in push/pull tandem pairs. Both these landplanes were dwarfed by the Dornier Do X flying boat.

Claudius Dornier was another of the German designers who, far from languishing in defeat after 1918, went from strength to strength, though at first it had to be outside Germany. His Wal (whale) flying boat, a development of wartime designs, first flew in November 1922. Noted for its slender hull, rectangular wing and tandem pair of engines on the centreline, it combined toughness, reliability and long range, and was made in an amazing variety of versions with different sizes of wing, different engines, tails and cockpits, and with weights from 8,818 to 22,046 lb (4,000 to 10,000 kg). When output finally ceased in the 1930s over 300 had been delivered, for all kinds of military and commercial purposes. The Wal was later sup-

Below: Flown by A. J. Cobham, this de Havilland D.H.50J had extra tankage and a hatch for a photographer. In 1925–26 it made a round trip to Cape Town.

plemented by the bigger Super Wal, most of which had two cabins for 21 passengers and four Jupiter engines. The Do X, however, was not a profitable venture, though technically adequate. First flown on 25 July 1929, this monster flying boat was by far the largest aeroplane of its day. Powered by 12 engines in six tandem pairs, it once took off with a crew of ten, 150 passengers and nine stowaways! Dornier sold two 'production Do X' flying boats to Italy, but none ever went into any regular airline service because about nine out of every ten seats would have been empty.

British airliners of the 1920s were wholly traditional in form, being fabric-covered biplanes held together with struts and bracing wires. The assortment of types derived from wartime designs were gradually replaced by specially built civil machines, and after the pioneer airlines had been merged into the national carrier Imperial Airways on 1 April 1924 there was at least assured stability in the industry. By 1926 the two mainliners were the Armstrong Whitworth Argosy for European use and the de Havilland DH.66 Hercules for Africa and routes to India, though passengers to these distant places had to travel across Europe by train, because of restrictions imposed by foreign governments. At first passengers had to take trains from Paris to Marseilles and then go to Port Said, Egypt, by steamer. As late as 1934 trains were still needed, with four changes, from Paris to Brindisi, where a biplane flying boat was hopefully waiting. The through trip from Croydon to Cape Town, even in 1934, was made up of 33 separate journeys in seven aircraft and three trains!

Despite such problems, the great global scheme of air routes very slowly took shape. Aerodromes progressed from being a field with a hut, notice board and windsock, to being recognisable as such with hangars, a central building with a control tower, probably a hotel, and a full range of lighting and radio aids. All mainline transports had two-way radio by the late 1920s, but the problem of fog, blizzards and other forms of bad weather remained unsolved. Airline captains became amazingly skilled, such as Capt R. H. McIntosh who in the years after World War I several times landed at Hounslow, or at London's new airport at Croydon, which was officially opened in March 1921, in fog so thick that car drivers on the roads gave up. 'All-weather Mac' as he was known, had nothing to guide him down to the ground except that watchers on the ground could tell him by radio when he was passing overhead. He then had to work out exactly how far to go in order to come back with the correct landing approach. Any of today's pilots would be terrified to be in such a predicament.

On top of all this, riding in a 90 mph (145 km/h) box held together with wires, wallowing through all the turbulence near the ground, was a certain way of inducing airsickness. Freezing cold, or alternatively local blasts of hot air mixed with exhaust fumes, severe vibration and deafening noise, were all thrown in as extras. Any passenger bold enough and rich enough to fly coast-to-coast across the United States got to know the crew pretty well especially as the co-pilot came back into the cabin to hand out box lunches and was statistically likely to find two out of every three passengers had been airsick. Steve Stimpson, San Francisco manager for Boeing Air Transport, suggested hiring young nurses. After three years spent getting nowhere, and watching BAT become merged with other lines into the mighty UAL (United Air Lines), he was suddenly told to hire eight stewardesses. Led by Ellen Church, the eight went into action on a Boeing 80A route, San Francisco to Chicago, in 1930. Their impact was considerable, and modern stewardesses who consider their life tough might ponder on the fact that, as well as coping with airsick passengers all round the clock, Ellen Church's stewardesses also clambered out on the wing at stopping places to do the refuelling.

Aeroplanes were still puny, but in the hands of skilled and dedicated pilots they conquered great distances and pioneered the air routes of today. After World War I many had expected flying to become the pastime of the ordinary person. Looking for fresh markets, manufacturers picked ultralight machines with only one or two seats, such as the Blackburn Sidecar with a 40-hp engine which was exhibited in London at Harrods store in March 1919. The universities of Oxford and Cambridge, whose undergraduates included many experienced ex-wartime pilots, fought a thrilling aerial counterpart to the Boat Race at Hendon on 16 July 1921, using six SE.5a fighting scouts! Cambridge won. The King's Cup air race was held annually from 1922 in Britain, while the Coupe Deutsch de la Meurthe in France and the Pulitzer Trophy in the USA became even greater attractions and, unlike the King's Cup, spurred the development of a host of exciting racers which were often technically ahead of the rather pedestrian fighters of the day.

Above: An unusual expression of entente cordiale *between British, French and Dutch airline pilots in 1926 Croydon. Centre is famed pilot 'all-weather Mac'.*

Even the relatively slow military machines made headlines. Right in the forefront was the US Army Air Service, though whittled to a tiny fraction of its 1918 size and starved of funds. In May 1923 Lts Kelly and Macready took off from New York in a single-engined Fokker T-2 and kept going for 26 hr 50 min until they landed at San Diego, the furthest point of the US they could reach. In 1924 Lt Russell Maughan boarded the newest pursuit (fighter), a Curtiss PW-8, at New York and despite having to stop for fuel reached San Francisco in under 18 hours, the first time

Left: The world's first air stewardesses pose before their United Boeing 80A in 1930: Leader Ellen Church stands third from left.

the continent had been crossed between dawn and dusk. Even bigger headlines were gained later in 1924 when two DWCs – standing for Douglas World Cruisers – completed the first circumnavigation of the globe. Four had started out from Seattle flying westwards via Alaska and Japan, and at times they changed their wheels for seaplane floats. Only two completed the gruelling journey on 28 September 1924, after covering 27,553 miles (44,342 km) in 175 days, the flying time being over 371 hours.

The conquest of the air was also demonstrated by other nations. As early as 1920 the Italian Air Force sent a formation of former wartime bombers, four Capronis and seven speedy Ansaldo SVAs, on a challenging expedition to Tokyo. Piloted by Ferrarin (lster a Schneider racer) and Masiero, two of the SVAs actually got there, but it took them three months. On 1924 Pelletier of France flew from Paris to Tokyo in a Breguet biplane bomber in 48 days despite a spate of hair-raising accidents, while in 1925 another Breguet was flown by Abe and Kawashi of Japan in the reverse direction with scarcely a hitch. Also in 1925 two Italians, the Marquis de Pinedo and Campanelli, flew a slow Savoia-Marchetti S.16ter flying boat on one of the longest voyages ever: 34,000 miles (54,720 km) from northern Italy, around Australia, up to Tokyo and back. Even the infant aircraft industry of the Soviet Union, barely recovered from years of civil war, managed to send a small group of aircraft which left Moscow on 10 June 1925 and reached Pekin on 17 July. Dornier Wals figured in several epic flights. Italy's Locatelli and crew failed in their 1924 attempt to be first to fly the Atlantic westbound, though the tough flying boat kept them comfortable until they were rescued. But a Spanish Wal, flown by Maj Franco and crew and powered by British Napier engines, crossed without a hitch in 1926 in a flight that linked Europe with Buenos Aires.

On 8 May 1927 one of the most famous flyers in the world, the great French wartime ace Nungesser, set off with fellow-officer François Coli in a Levasseur machine specially built to fly from Paris to Newfoundland, the PL.8 *Oiseau Blanc* (White Bird). They took off with high hopes, and were never heard of again. But less than two weeks later came the flight that received a rapturous reception around the world, and made one man more famous than any other pilot in history – certainly since the Wright brothers. He was Charles Lindbergh.

Just why this one flight so caught the imagination of the world is hard to fathom, though a few features in it were new. First, he did it alone and he did it with consummate skill. He got a special aeroplane built and planned every detail of the flight himself. He did it all with the glare of publicity on him and, incidentally, the US Press tended to be snide and sarcastic during his preparations, calling him 'The Flying Fool'. The flight did not start at a remote field and end in a forced landing in an even more remote place, as had Alcock and Brown

Below: The first aircraft to fly around the world were two (of four) DWCs (Douglas World Cruisers), ordered by the US Army Air Service especially for the task in 1923. Two of them completed the mission in 1924, using sometimes wheels, sometimes floats. This one, No 3 Boston, *ditched in the Atlantic.*

eight years previously; it started at a crowded aerodrome at New York and ended at an even more crowded one in Paris. Not least, Lindbergh flew solo, so that instead of the Flying Fool the headlines eventually called him 'The Lone Eagle'.

Born in 1902, tall and handsome 'Lindy' began as a daredevil barnstormer in 1920 doing wingwalking, parachuting and similar stunts before qualifying as a pilot in 1923. He flew so intensively that three years later he had logged 2,000 hours, most of it as a mail pilot between St Louis and Chicago. He was eager to do something big to demonstrate to a doubting yet optimistic public that the aeroplane had come to stay. His chance came when New York hotelier Raymond Orteig offered a $25,000 prize for the first non-stop flight from New York to Paris. Lindbergh got a group of St Louis businessmen to fund his attempt, so when his special Ryan monoplane had been built he named it *Spirit of St Louis*. Powered by a 220-hp Wright Whirlwind radial, the NYP (New York-Paris) was virtually a flying fuel tank housing 375 gallons (1,707 litres). Behind the main fuselage tank was the tiny cabin for the pilot, with a window on each side but with no forward visibility whatsoever.

Lindbergh was a naturally gifted pilot, an outstanding navigator and a meticulous planner. He attended to every detail he could – one of the last things he did before the great flight was to instal a special heater to cure severe carburettor icing – but he could not be sure that the engine would stay running for the more than 30 hours needed, and he was even less certain that he could stay awake. In fact his greatest mistake was to get no sleep on the night before his takeoff on 20 May 1927, so that he was tired before he started.

Somehow they got an overload of fuel into the NYP, and the little monoplane was

then so heavy that a crowd had to push on the landing gear and wing struts to start the ship moving on the rain-sodden grass of Roosevelt Field. Acceleration was so sluggish that many, including Lindbergh, felt 'He'll never make it!', but at last the overloaded NYP staggered into the air and just cleared the telephone wires at the far side of the field. Once airborne the challenge and excitement of the unprecedented solo flight tended to counter the tiredness and the soporific effect of the non-stop roar of the trusty engine. A little over 28 hours into the flight he crossed the Irish coast within three miles (4.8 km) of where he had intended, and the worst was over. Five hours later the second night had fallen, but after a search he found the Paris aerodrome of Le Bourget and made a dummy landing before committing himself to the final arrival. He still had no forward view and the field seemed to be thronged with the entire population of Paris, but he hit nobody.

He had covered the 3,614 miles (5,816 km) in the extremely good time of 33 hours 30 minutes. He could have gone on to Rome or beyond if necessary. As it was, not only had he hit the headlines in a way no other pilot in history has ever done, but he had beaten the new world non-stop distance record set up after a great struggle by two RAF men who had flown a Hawker Horsley torpedo bomber from Cranwell to Bandar Abbas in the Persian Gulf (3,420 miles, 5,504 km) over the very same three days, landing two hours earlier. This was tough on the Horsley's pilot, because nobody today has heard of F/Lt Carr, whereas everybody knows of Lindbergh.

Below: After his great flight 'The Lone Eagle' made a triumphant visit to Britain; here he is with the RAF at Gosport on 31 May 1927.

Though it was made of wood, the Lockheed Vega set entirely new standards in streamlining and in the speed of air travel, especially after it appeared with the Wasp engine. Varney, a Californian operator, later bought the even faster Lockheed Orion in 1931.

5. A GATHERING OF EAGLES

DURING THE 1920s AVIATION made great progress in outstanding pioneer flights over great distances, and in the setting up of struggling airline companies; while the determined efforts of the Bristol, Pratt & Whitney and Wright companies gave the world engines of increasing reliability which dramatically reduced the frequency of engine failures. But in most areas, including that of basic aeroplane design, progress was unimpressive. The best fighters, civil airliners and other machines of 1929 would have been instantly understood by any 1919 pilot, and though he would have been excited at the progress made, it was of degree rather than of any fundamental kind.

Yet in the background were a host of major advances which, worked on throughout the 1920s, were to transform the most powerful, high-speed aeroplanes of the 1930s to such a degree that half the fighters and bombers being built in 1937 were obsolete by the start of World War II. Some of these new developments were in aerodynamics, the most fundamental of all aeronautical sciences; others were in such fields as structures, propulsion and in the equipment carried on board to assist the pilot to fly in bad weather.

Before describing the great changes which transformed both the appearance and the performance of aeroplanes in the 1930s, it is worth studying the political scene. World War I was originally known as The Great War, and many called it 'the war to end wars'. Politicians in most countries naturally reflected popular opinion in wishing to do away with armaments such as military aircraft, and to hope that there would be no more war for many years. For example, in Britain the RAF was managed and funded on the basis of what was cosily called The Ten-year Rule, which simplistically assumed there would be no war for the next ten years. Thus in 1920 it was judged that no really new aircraft would be needed until 1930 and in 1921 the date was shifted to 1931! This procedure was hardly conducive to the introduction of new technology. But by 1930 crisis after crisis had loomed; during the 1930s Japan went to war with China and set up a puppet state in Manchuria; the Soviet Union went to war with Japan; Italy fought Albania and Abyssinia; Hitler seized power in Germany and announced the creation of the world's greatest armed forces; and a particularly bloody civil war erupted in Spain. All this darkened the previously bright skies and progressively opened the floodgates of government spending even in Britain, the nation where rearmament was resisted so long it nearly brought the country to its knees in 1940.

Aerodynamically the main problem in the early years of aviation was to achieve really good flight control. The subject was still in its infancy and the standards which pilots expected were extremely poor compared with those of today, but nevertheless the pilot had to be able to control the aeroplane. The standard form of cockpit flight controls was introduced by Esnault-Pelterie in 1907, with a pair of foot pedals for the rudder (pushing the left pedal to swing the nose to the left) and a stick for the ailerons and elevators (moving the stick to the left to roll the aircraft to the left, and pulling it back to raise the nose into a climb). All these motions were easily accomplished by the seated pilot and are natural in their sense. Little difficulty was experienced by pupil pilots in learning how to fly with such controls and, thanks to its introduction late in World War I, the Smith-Barry scheme of formalized pilot training had been adopted everywhere by the early 1920s. In the United States Kollsman perfected the basic mechanical or electromechanical kinds of flight instrument, while Elmer Sperry developed the autopilot so that aeroplanes thus equipped could be left to fly themselves for long periods. Even more important, so-called blind flying instruments enabled pilots to fly without needing to look outside the cockpit, and by the early 1930s military and commercial pilots were all proficient in flying 'under the hood', unable to see out. Thus, night and bad weather no longer held terrors, except for the still-unsolved problem of landing.

A few very large aircraft were difficult to fly simply on account of their size. Instead of a stick, such machines had an aileron control wheel which rotated with a gearwheel which drove a motorcycle-type chain connected to the cables driving the ailerons. The elevators and rudder were driven directly by cables. Often the length of cable driving the surface exceeded 100 ft (30 m), and there was high friction through the guiding pulleys and fairleads, quite apart from the fact that the large hinged surfaces needed great force to move them in themselves. Palliatives introduced by the 1930s included mass balance, horn balance (with the end of the hinged surface extended forwards of the hinge axis in what was called a horn) and inset hinges so that the entire leading edge of the surface was in front of the hinge axis. Some big machines used servo control surfaces, the pilot driving only a small auxiliary surface hinged on two arms well aft of the main surface, the latter then being moved by the aerodynamic load on the auxiliary 'servo' surface. By the 1930s the familiar 'normal' arrangement of parts for an aeroplane had become almost universal, so that so-called tail-first types were very rare, and not only did the monoplane gradually gain an ascendancy over the biplane but the biplane tail disappeared completely. The use of two or even three vertical tail surfaces, each comprising a fin and an attached rudder, remained as popular as ever.

In the 1920s there had been much argument about the best kind of wing profile, or 'aerofoil section', for each class of aircraft. Biplanes invariably had relatively thin wings as did all very high-speed aircraft such as fighters and racers. Heavy machines such as large flying boats, transports and bombers, had thick wings. In the case of large monoplanes the wings were so thick that their drag counteracted the advantages in 'clean' design (absence of struts and protuberances) compared with the biplane. In 1932 the British Air Ministry had two Blackburn twin-engined transports built as identical as possible except that one was a biplane and the other a monoplane. There was no difference in empty weight and virtually no difference in performance because the monoplane had a large, thick wing (span 86 ft/26.2 m, compared with 64 ft/19.5 m) braced by struts. But many aircraft designers realized by the 1930s that the monoplane could be made truly clean and that with modern structural methods its aerofoil section did not have to be so deep. This promised to give monoplanes a great advantage, not realized in the Blackburn because of its traditional structure.

One key to the development of the modern monoplane was thus aerodynamic, and the other structural. The aerodynamic key was to make the aerofoil section capable of being varied. The first important form of variation applicable to all aeroplanes was the Handley Page slat, introduced in 1919. In its original form the slat was like a very thin curved auxiliary wing, normally lying snugly against the leading edge of the main wing towards the tip. It could be opened by the pilot to stand away an inch or two from the main wing, leaving a slot through which the air would rush at increased speed. This kept the air flowing evenly back across the upper surface of the main wing and postponed the dreaded stall; this enabled the aircraft to be more heavily loaded or to fly more slowly, and thus (for example) to land in a smaller space. Later the slat was arranged so that, as the aircraft flew very slowly and neared the stall, the lift on the slat would pull it up away from

Right: Britain's Imperial Airways persisted in using 90-mph (145 km/h) fabric-covered biplane airliners until 1939. Here H.P.42 Horatius *sets course from Croydon.*

the surface of the main wing, making its operation automatic. In the 1930s Lockheed and a few other companies built the slots into the main wing, with hinged vanes to open them at high angle of attack. Today giant slats on jetliners are driven open hydraulically to keep colossal airflows harnessed and hundreds of tons in the sky.

The other aerodynamic key was the flap on the rear of the wing. Back in the first years of the century experimenters had toyed with the idea of a wing made in front and rear sections, hinged together so that camber could be varied at will. Gradually the flap matured as an auxiliary surface hinged or in some other way attached to the wing so that it could be tucked out of the way while in cruising flight, and for slow-speed flight it could be extended to increase both lift and drag. By 1930 several different forms of flap had been developed and during the 1930s these were applied to new designs of aeroplane. The simplest flap is the so-called plain type, where the trailing edge of the wing is hinged. The slotted flap is the same except that there is a slot between the wing and the flap, giving greater effect in increasing lift by keeping the air flowing across the flap's upper surface. The split flap cannot be seen from above the wing, because only the lower half of the rear part of the wing is hinged; this does little for lift, but greatly increases drag. There are many other varieties, but by far the most effective modern types are the Fowler flap and the double-slotted type. Harlan D. Fowler patented a flap running out to the rear under the fixed trailing edge on curved tracks; initial motion merely increases wing area, giving increased lift for takeoff, while extending the flaps fully also rotates them downwards to give very great lift and drag for landing. In the double-slotted flap there is a complete extra slat-like section carried on the front of the flap so that, when the flaps are fully down, there are two slots to keep the air flowing steeply down across the flap, which may be depressed to 50° or more.

Flaps enabled a given size of wing to give up to double the lift at low speeds; in other words, it became possible to think in terms of much greater wing-loading. Wing-loading is the weight supported by each unit area of wing, in Imperial units measured in pounds per square foot. In the 1920s wing-loadings gradually crept up from around 10 lb/sq ft to 20 lb/sq ft, but with the advent of streamlined monoplanes with powerful flaps on the wings in the 1930s the figure leaped up to 30, 40 or 50 lb/sq ft. Thus in cruising flight the wing could be small and streamlined offering little drag, but for landing the flaps were extended, and possibly slats opened, transforming the aircraft into a good slow-flying machine able to land in the same short space as the lightly loaded biplanes. Indeed, by 1942 the B-29 bomber was flying with its wings loaded to an awesome 78 lb/sq ft, which a few years earlier would have terrified pilots and been incompatible with aerodromes.

At this point it is essential to comment that the aerodromes did not themselves stay quite the same. Croydon aerodrome, London's new airport in 1921, was in effect a grass square 1,500 ft (457 m) across. By the late 1920s paved runways were common in the USA, with lengths generally greater than the best run available at Croydon. Full electric light installations round the boundaries of major airports were installed gradually, together with rows of lights showing suitable landing directions. In 1937 the new airport for Prague at Ruzyne, was built in a field with a maximum dimension of 3,281 ft (1 km) across, with four lines of lights spaced every 45° to give eight takeoff or landing directions. But in the same year the much smaller city of Stavanger, in Norway, opened a new airport at Sola with two concrete runways, each almost 3,281 ft

Above: When Croydon opened in 1920 passengers were handled in tents and huts. A new building was constructed serving as booking hall, and control tower.

(1 km) long and 128 ft (39 m) wide. Such runways were gradually realized to be an essential requirement for the safe all-weather year-round operation of the new generation of high-performance airliners that were arriving in the mid-1930s.

Having described the aerodynamic advances, the structural changes can be fairly said to have been of equal importance. It will be recalled that until 1930 most aeroplanes were made in the form of a wire-braced skeleton covered with fabric. The skin was strong enough to provide lift and sheathe the skeleton in a streamline form, but not to support any structural loads. The skeleton was variously of wood, or of beams built up from rolled sections of very thin steel joined by rivets or other attachments, or from regular steel tube cut and bent to shape and welded together. Welded tube was generally the cheapest method, while wood was easiest for most workers and factories and certainly easiest to repair. The only exception to this form of construction was the so-called semi-monocoque method, in which a substantial part of the load was borne by the skin. This began in 1911 with streamlined monoplanes designed by Louis Bechereau, including the Deperdussin racers, whose beautifully shaped fuselages were formed from thin layers of veneer arranged one over the other. Other makers, such as Dornier, Short and Rohrbach, sought to do the same in metal.

The very first metal-skinned aircraft, such as Reissner's of 1912, used thin sheet steel, but it was too heavy. By 1918 the new alloy of aluminium, copper and other elements called duralumin was becoming available in quantity, and this offered the low density of aluminium combined with much higher strength. Junkers used it for both the underlying skeleton and the corrugated skin. It is especially worth noting the way his monoplane wings were built with as many as five or six tubular spars along the top and as many more along the bottom, all joined by numerous bracing struts, finally to have the corrugated sheets wrapped round the leading edge and joined further back. Junkers opened a factory in the Soviet Union, and here such designers as Tupolev, Petlyakov and Myasishchyev all worked on the same team to develop this form of construction further. In 1925 the Tupolev ANT-4 took it to a cantilever monoplane aircraft weighing over 15,000 lb (6,804 kg) and powered by two 500-hp engines. In December 1930 came the ANT-6, the world's first modern-style four-engined monoplane with four 500-hp engines, weighing up to 53,000 lb (24,040 kg). In 1934 the mighty ANT-20 *Maksim Gorkii* took to the sky with eight 900-hp engines and weighing 92,593 lb (42,000 kg). In 1936 the Soviet authorities reluctantly stopped construction of the ANT-26, a gigantic bomber with 12 engines of 900 hp, a span of 311 ft 8¼ in (95 m) and a weight of 167,500 lb (76,000 kg). All these used refined versions of the Junkers form of construction, but by 1936 not only did this seem conceptually dated but the idea of a gigantic bomber, which might possibly be destroyed by a direct hit by a single large AA shell, appeared a poorer investment than 20 or 30 speedy twin-engined machines costing the same amount of money.

Junkers himself reached the pinnacle of his traditional corrugated-skin machines with the relatively modest Ju 52 of 1930, with a span of 95 ft 11½ in (29.25 m) and a

Below: Best-selling European airliner of the 1930s, the Ju 52/3m was also the Luftwaffe workhorse in World War II.

Left: British Airways bought the Lockheed Electra in 1936. It was one-third as fast again as any available British airliner.

single engine. Only five of these were built, but in May 1932 the three-engined Ju 52/3m arrived and swiftly became the most important civil transport and derived aircraft of the entire 1930s. The German state airline Deutsche Luft Hansa alone used over 120, and total production at the Dessau factory was 563 prior to the outbreak of World War II. Production of this corrugated-skin 17-seater, typically with three 830-hp engines, was then stepped up and it served as almost the standard Luftwaffe transport throughout the war. Including production by Amiot in France for the Luftwaffe, the total number built was 4,845 by May 1945. The French then made another 400 post-war, 170 were built in Spain and a handful survived in Switzerland until the late 1970s. The great attributes of the *Tante Ju* (Auntie Ju) were toughness, reliability and an amazing ability to get in and out of very small, rough fields.

At the same time, while it was an excellent passenger liner in 1932, it was clearly not the best that could be done in World War II. It was small, so that there was no room to stand upright in the cabin, nor have more than one row of seats under the windows on each side, with a narrow central aisle. The floor sloped when on the ground and everything had to be loaded through a small side door. It was difficult to get a fuel or oil drum or motorcycle aboard and anything bigger could not be carried at all. The next chapter describes how Hitler tried to provide bigger-scale airlift for his vast armies.

One of the few rivals to Junkers in providing transports for DLH (Deutsche Luft Hansa) was the versatile firm of Ernst Heinkel. This mercurial character startled everyone in December 1932 with the He 70, a trim low-wing monoplane. It looked like the sort of thing 1932 schoolboys might have drawn as a 'ship' of the future, with cantilever elliptical wings, an almost perfectly streamlined duralumin fuselage and, a novel feature for 1932, retractable landing gear folding outwards into the wings. A few racers had tucked away their wheels to reduce drag as early as 1920, the first being the Verville-Sperry and the Bristol 120. Today we are so used to neat retracting gears that it seems strange that in the early 1930s designers should have found it so difficult, using extremely complex and clumsy arrangements which caused prolonged difficulties and frequently failed to function reliably. On top of this, it was then common for the pilot to forget to lower the wheels prior to landing, even after designers had added a loud alarm bell, klaxon horn or flashing light to give warning whenever the throttle was closed with the gear still selected in the 'up' position.

At least the He 70 had a neat arrangement, rather like that used later on the Bf 109, Spitfire and some other fighters, though at least one Luft Hansa line pilot forgot his gear and did a perfect belly landing with passengers aboard. From the economic point of view the He 70 was not attractive, because it only carried four passengers, but had a thirsty 750-hp engine. It was built purely for speed, and the reason for its appearance was that Swissair had placed in service on the highly

Below: This Lockheed Vega was one of the most famous aircraft of all time. It was twice flown round the world by one-eyed Wiley Post – the second time, in July 1933.

competitive Zürich-Munich-Vienna route a pair of American-built Lockheed Orions. At that time US-built aircraft were very rare in Europe, and the Orion created a sensation. Among other things, it was roughly twice as fast as most airliners in European service.

The Loughead (pronounced Lockheed) brothers had built a few aircraft in World War I, but it was not until 1926 that the Lockheed company was started. John K. Northrop designed their first major product, the Lockheed Vega, which though made of wood was probably the most streamlined transport of its day. After the first flight on 4 July 1927 the test pilot, Eddie Bellande, said "You'll sell this airplane like hot cakes". One of its features was that the curving fuselage was exactly tailored for minimum drag behind the single radial engine, and Vegas flew with various engines, starting with a 220-hp Whirlwind like that used by Lindbergh, completely uncowled, and ending with a 500-hp Wasp with the installation surrounded by a long-chord cowl. The installation of radial engines made great progress in the early 1930s, the first common type of cowling being the Townend ring, which was like a small rectangular wing bent round into a ring around the cylinders to give extra thrust. By the mid-1930s cowlings had become longer, enclosing the engine completely. Once cooling problems had been overcome, the fully cowled radial usually gave less drag than the seemingly more streamlined liquid-cooled type of engine and was invariably lighter and cheaper, quite apart from being far superior in both very cold and very hot climates.

Below: The sole DC-1, pictured on its first flight on 1 July 1933. Because of a carburettor problem, the engines stopped whenever the DC-1 tried to climb!

Altogether Lockheed sold 141 Vegas, the later models being designed by Gerard Vultee who like Northrop later had his own company. Many were used by commercial airlines, together with a slightly different model called the Air Express, with an open pilot cockpit and the wing mounted above the fuselage. By far the most famous Vega was named *Winnie Mae*, which today is a star exhibit in Washington's National Air and Space Museum. The trim white Vega was flown around the world in June/July 1931 by the famed one-eyed pilot Wiley Post, with Harold Gatty as second crewman. Far more amazing as an accomplishment is that in 1933 Post not only flew round the world again in *Winnie Mae* but he did it solo, in the record time of 7 days 18 hours 49½ minutes, virtually without sleep.

The Orion, which so amazed the Europeans, was in effect great-grandson of the Vega. The son was a series of low-wing racers of 1929–30 called Explorers, an offshoot of which was the Sirius used by the Lindberghs (Charles and his wife Anne Morrow Lindbergh) on many record-breaking flights, including non-stop California to New York in 14 hours 45 minutes on 20 April 1930. The Lindberghs continued flying intensively through the 1930s, much of it surveying future routes for Pan American. The Sirius led to the Altair ultra-fast mailplane which was powered by Wasp or Cyclone engines of up to 650 hp; one was used for record-breaking flights by Charles Kingsford Smith. Last of this series were the Orion passenger transports, 35 of which were built. Most had six passenger seats, and cruising speed varied from 175 to a possible 200 mph (322 km/h), making possible an 86-minute schedule between Los Angeles and San Francisco when the

fastest alternative was $2\frac{1}{2}$ hours.

Sadly, Lockheed became part of a giant conglomerate which milked all the profits, causing bankruptcy. In stepped young Bob Gross, a business graduate from Harvard, who bought the assets on 6 June 1932, at the height of the great Depression. The US judge clearly thought him insane, and asked "I hope you know what you are doing?" Gross did, though for the next two years the reborn Lockheed was mortgaged up to the hilt. When the brand-new Electra twin-engined airliner jammed its landing gear in the retracted position in 1934 it looked like the end, but test pilot Marshall Headle pulled off a masterly landing with minimal damage. Soon Lockheed sold 148 of the trim all-metal Electras, most of them seating ten passengers. From them stemmed the smaller Model 12 and the larger and much more powerful Model 14, the first production aircraft with the highly effective Fowler flaps, which carried 12 passengers at an easy 225 mph (362 km/h), faster than any other airliner in the 1930s. One took Prime Minister Chamberlain to Munich in 1938, and in the same year Howard Hughes flew one round the world in the staggering time of 3 days 19 hours 8 minutes.

These speedy Lockheeds, all with twin-finned tails, went for performance rather than capacity, but they had competition from two other companies which, up to the present day, have dominated the commercial transport market around the world: Boeing and Douglas. Boeing himself carried the first bag of international air mail, flown in a Boeing Model C from Victoria BC to Seattle on 3 March 1919. In 1927 Boeing applied to operate the mail route from Chicago to San Francisco and built a fleet of Model 40A mailplanes with the new Wasp engine. Whilst churning out scores of fighters for the US Army and Navy, Boeing also built the big three-engined Model 80 biplane liners used by the world's first stewardesses, as related in Chapter 4. The Model 80 family had such unknown refinements as hot running water in the hand-basin and an electric light for each seat.

Boeing's Model 80 family had an immediate rival in the Curtiss-Wright Condor, another big biplane which began life as a US Army bomber in 1928 and developed into two quite different generations of civil transports in the 1930s. The final batches had single-fin tails, large fuselages with room for sleeping accommodation, and 750-hp Cyclone engines into whose nacelles the main landing gears could retract, giving an odd mixture of old and new (there were not many biplanes with retractable landing gear). Both Boeing and Curtiss recognised that all these biplane liners were basically outdated, aerodynamically and structurally.

As explained earlier, the ideal form of structure was known to be the monocoque form, like the claw of a lobster, with all the strength in the skin and no need for an internal skeleton. The wooden semi-monocoques were well-developed by 1930, the Vega being an example, but for airline use such aircraft could too easily be damaged by impacts from ground vehicles, ladders and other hard objects. The use of stressed-skin construction using duralumin could have made rapid progress in the 1920s, but it was confined to just a few little-known machines, notably the landplane and flying-boat transports designed by Dr Adolf K.

Below: 'Peace in our time' proclaims Prime Minister Chamberlain on his return from Munich in September 1938. He flew in G-AFGN of British Airways.

Left: The DC-2, powered by 710-hp Cyclone engines, was the development of the DC-1. It was a great success, but as nothing compared with its larger successor, the DC-3.

Rohrbach. His Zeppelin Staaken E.4250, a high-wing four-engined airliner first flown in 1920, was a truly futuristic machine and showed how airliners would be built in the 1930s. The cantilever wing had as its structural basis a very strong box-like structure formed by the curved top and bottom skins and the full-depth spars, all made of strong duralumin. On to this were added the light, almost unstressed leading

Below: Though British firms failed to build a modern airliner, they did build a long-distance racer in the D.H.88 Comet. This particular Comet won the 1934 MacRobertson race to Melbourne. It still exists and by late 1984 had been completely restored to flying condition.

and trailing edges. Such wings were seen in improved form in Rohrbach's Roland landplanes and Romar flying boats, which were used by Luft Hansa in the late 1920s. Their wings would not look out of date today.

In 1927 Rohrbach gave a lecture in the USA and this triggered off the Boeing management and also John K. Northrop. The latter got hardware flying first, with his Alpha of spring 1930. This was similar to the fast Lockheed low-wing machines except that it had all-metal stressed-skin construction and fixed landing gear. It led to the Gamma and Delta, all high-speed mail and passenger machines of the early 1930s, whose capabilities were highlighted by a flight made by the famed Jack Frye, vice-president and chief operating manager of TWA, which crossed non-stop from Los Angeles to Newark, New Jersey, in 11 hours 31 minutes in May 1930. As for Boeing, this company flew its Monomail only a few days after the maiden flight of the Northrop Alpha in May 1930. This combined stressed-skin construction with retractable landing gear, although the pilot still had an open cockpit (as he did also in the Alpha, because pilots did not like the idea of an enclosed cockpit which might spoil the view).

Boeing also built the B-9, a twin-engined bomber, and this greatly assisted the design of the next commercial transport, the Model 247. This is often called 'the first modern airliner', although the original 247 had few features that were not in the German E.4250 back in 1920. However, it did have a low wing and retractable landing gear, the wheels folding back into the nacelles of the 550-hp Wasp engines, which had Townend-ring cowls and drove propellers with three metal blades. The fuselage, of stressed-skin construction with a smooth exterior like the wing, was rather spoilt aesthetically by the forwards rake of the windshield for the pilot. This was com-

mon in the early 1930s and was a feature of the prototype Lockheed Electra; it was thought to give better view ahead for landing. Down each side of the main cabin were five comfortable armchair seats, each next to a wide window made of celluloid, which was then invariably used instead of glass in aeroplanes because it was much lighter. The seat pitch of 40 in (just over 1 m) seems princely compared with the 29 in (737 mm) that is common today, but on the other hand, as in most early low-wing airliners, the main wing spars passed through the cabin and passengers had to step right over them.

The 247 first flew on 8 February 1933 and the United airline group ordered an unprecedented fleet of 59. This was the biggest airline order in history and appeared to be enormously good fortune for Boeing. In fact it was quite the reverse. Jack Frye of TWA wanted a new transport to beat the forthcoming Boeing and issued a specification suggesting three engines carrying 12 passengers at 150 mph (241 km/h). Douglas Aircraft had bought Northrop's company and was eager to produce a modern stressed-skin machine, and it badly needed the work. It calculated it could do better with two of the modern engines then becoming available, such as the Hornet or Cyclone, than with the old three-engine formula. Moreover, there were yet other new advances waiting to be used. One was wing flaps, to enable a small streamlined wing to lift a heavy aeroplane at low speeds on takeoff or for landing. Another was the long-chord engine cowl, to reduce drag without harming cooling. A third was the variable-pitch propeller. The old fixed-pitch propellers were ideal only at one speed; it was rather like driving a car with only one gear. This did not matter very much when aircraft did not exceed 100 mph (161 km/h), but by the 1930s speeds were already double this, and it was realised that what was wanted was a propeller that could be set to fine pitch for takeoff or landing, so that the engine could develop full power at high rotational rpm even with the aircraft hardly moving, and which could progressively be set to coarse pitch in cruising flight so that, on each revolution of the propeller, the aircraft would travel a long distance forward. The American Hamilton company pioneered two-pitch propellers by 1931 and by 1934 were in a position to offer so-called constant-speed propellers, the rpm of which could be set to any desired level according to the engine power needed. An automatic governor then adjusted the pitch of the blades according to whether the aircraft was beginning its takeoff run, climbing, cruising, diving or landing.

TWA accepted the Douglas DC-1 (Douglas Commercial type 1) and the aircraft flew on 1 July 1933. But the Boeing 247 had entered service with rival United on 30 March 1933, and proved so hurtful in competition that TWA asked Boeing to sell it a fleet of 247s after the first 20 had gone to United. In one of the business decisions of history not recognised at the time as historic, Boeing refused; its primary allegiance was to United with which it was then affiliated. Accordingly, in August 1933 Frye ordered from Douglas 20 of an improved DC-1 version called the DC-2. This was to launch the DC family of transports, which as early as 1936 were able to sweep away virtually all competition on the world's mainline routes.

Below: Armstrong Whitworth designed a modern airliner in the Ensign, first flown in January 1938, but it was obsolescent in concept and also had indifferent engines.

Above: Parked at Clover Field, Los Angeles, before its first flight on 17 December 1935, the DST (Douglas Sleeper Transport) was the harbinger of the future.

The DC-2, which flew on 11 May 1934, was slightly larger than the DC-1, and it had seven instead of six seats down each side of the cabin. Its smooth, stressed-skin construction contained a feature that nobody recognised at the time as important. Northrop favoured the use of multiple wing spars and he put three into the new Douglas in order to spread out the heavy bending and shear loads, thus reducing the weight of the wing. Right from the start it was found that this structure was well suited to small changes to allow it to bear much greater loads, so that the same basic wing could be used for the DC-1 weighing 17,500 lb (7,938 kg), for the DC-2 at 18,080 lb (8,201 kg) and the later DC-3 at 24,000 lb (10,886 kg). Nor was this all. Prior to the 1930s few aircraft of any type ever flew as much as 1,000 hours, but after World War II airliners had to fly more hours a day and for longer lives (up to 40 years with modern transports). Thus fatigue became important. Fatigue is the sudden snapping of a piece of metal after it has been bent to and fro many times. With old aircraft it was not very important, but today it is absolutely vital to design aircraft so that even after thousands of flights, often in turbulent air which flexes the wings up and down, there will be no crack or, if a crack does appear, it will be noticed and can be repaired without the wing coming off. British airliners after World War II, such as the Viking, Marathon, Prince and Dove, had single wing spars. If they cracked, the wing broke. In contrast, the DC-1 set the Douglas Commercials off from the start – by sheer chance – with wings good for 50 or 60 years. Many of these machines have flown 60,000 hours, and no wing has ever come off.

In Britain the appearance of such fast modern liners was shrugged off. They were far away and even when they began to penetrate British markets no action was taken – except by newspaper magnate Lord Rothermere, who was so concerned at the situation he ordered from the Bristol Aeroplane company a modern aeroplane like the new American machines, but only the one was built. Imperial Airways continued to fly giant biplane landplanes and flying boats without cowlings, stressed-skin, flaps, variable-pitch propellers, retractable landing gear or any other new feature, cruising at a stately 90 mph (145 km/h). Then in October 1934 Britain played host to a galaxy of aircraft eager to win the greatest-ever intercontinental air race: the 'MacRobertson race' from Mildenhall to Melbourne.

The de Havilland Aircraft company, which had lately moved from Edgware in North London to Hatfield, was famed for its popular series of Moth light biplanes, powered in their later versions by the company's own four-cylinder aircooled Gipsy engines. The company could see that American aircraft would walk off with all the main prizes unless something was done, so it courageously decided to build a new long-distance racer, the DH.88 Comet. Made of wood, it was beautifully streamlined, and was powered by two of the specially made six-cylinder versions of the Gipsy engine, giving 205 hp each. This was enough for two pilots in tandem in an enclosed cockpit and a heavy load of fuel, and for the first time in Britain features included flaps, retractable gear and variable-pitch propellers. No such propellers were avilable, and the first set were obtained from the Ratier company in France. The de Havilland firm lost on the programme, because only three were sold, but the honour of the country was upheld in that one machine – G-ACSS, the scarlet *Grosvenor House* (the customer was the famous London hotel's managing director) flown by C. W. A. Scott and T. Campbell Black – won the race in a time of 70 hours 54 minutes. But the next to reach Melbourne was a DC-2 airliner of KLM, carrying both passengers and mail, and this swept the board in the handicap section. The presence of the shining DC-2 on the grass at the starting line did much to jolt the British designers and customers into the new era.

Imperial Airways took a decision in summer 1934 that profoundly affected trunk routes throughout the British Empire. Realizing that these very long routes were served by poor airfields, the decision was taken to go over almost entirely to flying boats instead of bringing the fields up to the status of modern airports. Short Brothers supplied a great fleet of 28 S.23 Empire flying boats, with modern stressed-skin structures and powered by four 920-hp Bristol Pegasus engines. Later the order was changed to 25, plus six for Qantas of Australia to handle the far end of that particular route, followed by nine S.30s with much greater fuel capacity and the new Bristol Perseus sleeve-valve engine, four of which were equipped with the first operationally effective form of inflight refuelling for non-stop operation on the North Atlantic. In themselves these were all excellent aircraft, but their use dramatically increased costs because, even ignoring the much higher overheads, their direct cost per seat-mile was more than double that of the rival Douglas landplanes, which were preferred by all the other airlines. Moreover they meant that Britain and her Empire entered World War II with the vast job of improving airfields still waiting to be done. To partner the flying boats Imperial did order a modern landplane, the Armstrong Whitworth Ensign, but this ran three years late (the fourteenth and last was four years late) and was both unreliable and totally uncompetitive.

Tragically, such experiences led to Britain's national airline, renamed BOAC on the outbreak of war, having a strong bias in favour of American aircraft. Boeing had tried to counter DC-2 competition with the improved 247D, with fully cowled engines driving through gearboxes to controllable-pitch propellers. The year 1934 had seen 26 scheduled transports in the USA forced down by carburettor icing alone, and ice accretion on wings and tail caused several fatal crashes. Now that services were being flown in bad weather and in cloud it was essential to combat this danger, and the 247D had carburettor heaters and de-icing overshoes on the leading edges of wings and tail. These overshoes, or 'boots', were long flexible tubes, normally lying flat; when inflated at intervals by compressed air the ice on them broke up and blew away.

Boeing failed to compete even with the DC-2, and meanwhile Douglas looked ahead to an enlarged DC-2 to meet the needs of American Airlines for a DST (Douglas sleeper transport). To replace the old Condor biplanes American needed a machine with 14 bunks, folded out of the way by day above normal seats. This meant a

Below: Mitchell's Supermarine S.4, built for the 1925 Schneider Trophy race, showed his flair for streamlining. A decade later he designed the famed Spitfire fighter.

wider fuselage than the DC-2 and increased weights, but this was all to the good as it gave higher structural and aerodynamic efficiency. As powerful new Cyclone and Twin Wasp engines of 900 to 1,200 hp were coming on to the market – the Twin Wasp was first used on the giant Martin 130 *China Clipper* flying boat used on transpacific service by PanAm – Douglas went ahead and flew the DST on 17 December 1935. Its new pointed wingtips and better streamlining resulted in drag being no greater than that of the smaller and lighter DC-2, and it cruised much faster and had longer range. The DST entered service on 11 July 1936, but by this time what had been called the Dayplane, given the designation DC-3, was becoming more important. As a day airliner it carried 21 passengers, seated 2-and-1 in line with the windows instead of the DC-2's 1-and-1. Gradually the word spread that the DC-3 was a real winner and it quickly became the first airliner to be a world standard type. Even the Soviet Union and Japan bought licences to build it, and deliveries reached 150 by 1 January 1938 and over 800 by 1 January 1942, excluding the licensees. In World War II militarised versions became the standard workhorse transport and glider tug of the Allies, and 10,926 were built. Hundreds are flying today.

It is an indictment of failure to keep abreast of the technology to note that in 1936 a DC-3 could carry 21 passengers at a speed that would outrun any fighter in the RAF, fly over 2,000 miles and go on doing it for 70,000 hours. The old-fashioned machines were simply not in the same class, particularly in such matters as reliability and longevity. The average flight time on RAF aircraft scrapped in 1931 was an appalling total of 81 hours. At that time the RAF had already spent five years searching for a modern fighter to break away from World War I technology, but it had little success. It finally picked the Gladiator biplane, the only new feature of which was to have four small machine guns instead of two and which did not enter service until 1937. Fortunately, S/L Ralph Sorley, a junior officer, had calculated that in a future war eight such guns would be needed, while the Rolls-Royce company had in 1934 run the first of its Merlin engines.

Rolls-Royce gained a great boost by the Schneider Trophy contests, which taught the company how to wring more power out of their style of liquid-cooled V-12 engine. Thanks to Fedden's drive at Bristol, the rival aircooled Pegasus and Mercury had far surpassed all other engines in powering aircraft for the RAF and the airlines, and in the long term were to lead to later radials used in all the last species of high-speed fighters after World War II, but in 1934 it looked as if the liquid-cooled V-12 was superior. Accordingly the Merlin was picked for completely new designs of fighter by Sydney Camm at Hawker Aircraft and Reginald Mitchell at Supermarine. Camm produced the Hurricane, first flown in September 1935 and a cross between new and old, with complex wire-braced structure and fabric covering and an old-fashioned wooden propeller, but armed with eight machine guns and fitted with flaps and retractable gear. The Spitfire was a modern stressed-skin machine, utterly unlike Mitchell's Schneider seaplanes, though it had many structural features (such as an elliptical wing with a single main spar built

Below: Two Gladiators of RAF No 73 Sqn, newly delivered in late 1938, represented the last of the agile biplanes. The war showed that the biplane concept was outdated.

up from a series of square tubes nesting one inside the other) which made it unnecessarily difficult to make. Both machines had their coolant radiators far from the engine, the Hurricane's being under the belly and the Spitfire's under the left wing, with a small oil cooler under the right wing. From the start the nimble Supermarine fighter was a delight to fly, but it took a long time to get production going and war clouds were gathering. So ominous was the situation the directors of the Hawker company took the bold and patriotic decision to build a new factory at Langley, near Slough, and to go into production with 1,000 Hurricanes even though the Air Ministry had failed to order them.

Such a decision was welcome, because the need for a stronger RAF was becoming more urgent by the hour. In many parts of the world there were recurrent crises, if not open warfare as in South America and China. In 1933 Hitler seized power in Germany and set about openly building up limitless armed forces spearheaded by the new air force, the Luftwaffe. The United States was strongly isolationist; it was not only militarily weak but was also disinterested in getting involved in the conflicts of others. France, the strongest airpower in the early 1930s, was finding it impossible to produce an engine to rival the Merlin and impossible to create good designs of combat aircraft; it had an industry fragmented into hundreds of companies which were in chaos by 1936 because the socialist government had ordered them to be nationalized into geographically arranged groups. This almost completely thwarted the once powerful nation's efforts to build up its Armée de l'Air into a modern fighting force in its hour of need, so that by 1938 panic orders on an enormous scale were being placed with the USA, Netherlands and other countries. Few of the total of more than 4,000 machines thus ordered ever reached France, which collapsed in June 1940.

Germany rapidly built up its air force until it was the world leader in 1939 in both quantity and quality of hardware. Its early combat machines were unexceptional, and the first heavy bombers – the Dornier Do 11 and Do 23 – were poor by any standard. While it was popularly thought in other countries that Hitler was building bombers in the guise of high-speed civil transports, the reverse was true of the Dornier Do 17 – called The Flying Pencil because of its slenderness – which was rejected by Luft Hansa as a very cramped passenger carrier and languished in a hangar until it was tested by Robert

Untucht, a captain in the airline and an officer in the RLM (Air Ministry), who decided it would make a fine bomber. More important was the Heinkel He 111, first flown a few months after the Do 17 in February 1935. This had an enormous elliptical wing and resembled a scaled-up twin-engined He 70, with a beautifully streamlined fuselage, It was genuinely planned as an airliner as well as a bomber, and six were used on Luft Hansa prestige routes, but its main role was that of a bomber. Both Fascist nations, Germany and Italy, eagerly seized on the civil war that erupted in Spain in July 1936 as the ideal testing-ground for their new and important warplanes. This gave both nations a very big advantage not only in ironing out operational shortcomings in their aircraft but in providing a vital core of experienced pilots who had learned in the best possible way how to deploy and manoeuvre fighters and bombers in modern warfare. The He 111B got into action in Spain in 1937, and proved so much faster than most of the opposing fighters that its defensive armament of three hand-held machine guns seemed entirely adequate. In 1940 this belief was to be rudely shaken.

Germany's chief fighter materialised in an unusual way. Prof Willy Messerschmitt was noted for light sporting machines and had no success trying to get orders from the RLM. In desperation he turned to exports and built the M 36 transport for Romania, receiving a blast of criticism from the RLM for his lack of patriotism. He explained the situation publicly, whereupon the RLM was almost obliged to place a contract with his BFW company, which later led to the Bf 108 Taifun, a four-seat cabin sporting machine of stressed-skin construction; it was so modern that it differs in no important respect from similar machines of today. Then, in December 1933, the Luftwaffe needed a new fighter, and the BFW firm was given a copy of the specification, although because of their political opposition few of the Nazi officials thought the resulting prototype would be worth considering. Indeed, when Ernst Udet, famed World War I ace and the head of Luftwaffe procurement, saw the resulting Bf109 he exclaimed to Messerschmitt "That thing will never make a fighter!"

At the time many fighter pilots insisted on flying an open-cockpit biplane, for the best visibility and the best manoeuvrability, and this view was to persist in Italy well into World War II. Even such modern attributes as retractable landing gear were looked on with disfavour. In May 1935, on a field outside Augsburg, stood the prototype Bf 109, a slim rakish low-wing machine, with stressed-skin construction, an extremely small cantilever wing with large slats as well as flaps, and a narrow cockpit enclosed by a hinged transparent hood. The main landing gears were pivoted to the fuselage and folded outwards into the

The RAF gets its first look at the Supermarine Type 300, about to be named Spitfire. Designer Mitchell holds his hat. Another 22,889 were to follow!

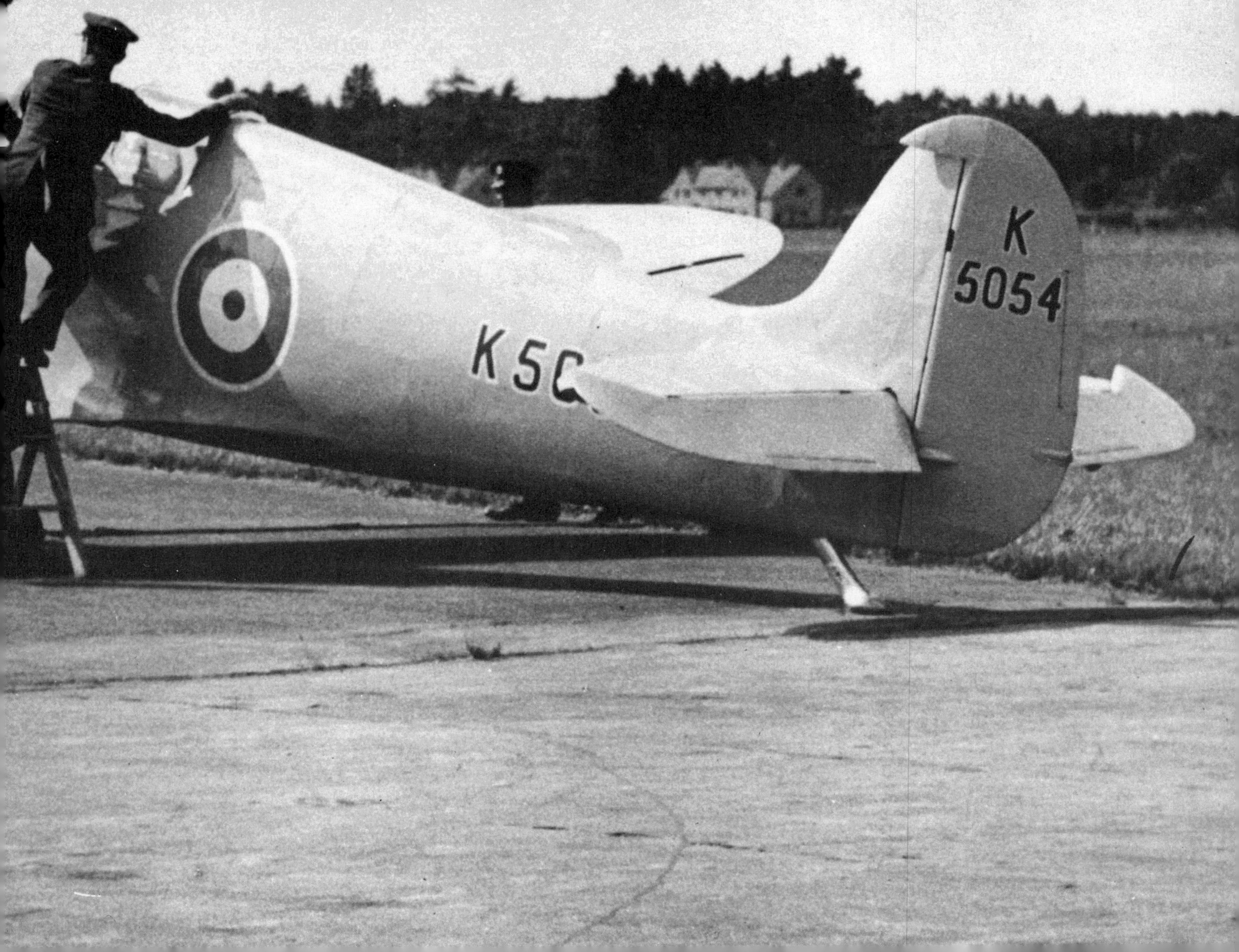

wings. At first this machine, which to 1934 eyes looked more like a racer than a fighter, received a barrage of politically inspired criticism. Heinkel's He 112 was the obvious favourite, but over the next 18 months the Bf 109 eroded all opposition by sheer merit and emerged as the best all-round fighter in the world. The fourth, fifth and sixth prototypes were hurriedly crated up in January 1937 and sent to Spain, where they quickly proved themselves far superior to all the motley collection of warplanes fielded by the Republicans. Thus the stage was set for ten years of non-stop production which grew to a fantastic crescendo of 14,112 Bf 109s in the year 1944, to a final (unknown) total in the region of 33,000.

The reason the new Messerschmitt was sent to Spain was because none of the Nationalist biplane fighters, including the German He 51, could stand up to the Polikarpov I-15 biplane and I-16 monoplane sent to the Republicans from the Soviet Union. The debate there over agile biplanes versus speedy monoplane fighters had been prolonged, and was not resolved by warfare against Japan over Manchuria and other territories. In Japan even the monoplanes were designed for manoeuvrability at any price, and were very lightly built by Western standards. In fact, the stumpy I-16 of December 1933 was the first low-wing monoplane fighter with retractable gear to fly, but it was tricky to handle, being almost unstable in some conditions, and was no match for the German 109, either in 1937 or when the two met in earnest in August 1941.

Fighting in Spain had shown that the Italian Fiat C.R.30 and 32, both traditional biplane fighters with open cockpits and fabric covering, were among the most agile aircraft in the sky. Fitted with large Fiat water-cooled engines, they were exceptional in both their climb and dive performance and rate of turn, and a few pilots amassed impressive scores whilst flying them. This confirmed the Italians in their view that such fighters were not yet outmoded. And the prowess of Italy in the Schneider Trophy races and on the Grand Prix motor racing circuits was reinforced by the brilliant aerobatic performance of the Regia Aeronautica's special teams – and not least by the fabulous transatlantic flights by General (later Marshal) Italo Balbo and his formations of large twin-hulled Savoia-Marchetti S.M.55 flying boats. The first 'Balbo' (the word came to mean any large formation) was of 12 S.M.55s which flew to Brazil in January 1931. On 1 to 15 July 1933 he led 24 of the big 'boats to the Century of Progress Exposition at Chicago, a truly remarkable achievement. French fighters detailed to escort the 'boats across France were embarrassed to find they had difficulty in keeping up, even at full throttle! Italy's reputation as an air power stood high in the 1930s, although politically the nation lost support for its aggression against Albania and Abyssinia; it used poison gas in Abyssinia, dispensed from its lumbering Caproni bomber-transports which of course met no air opposition.

In the United States, pursuits (fighters)

Despite its open cockpits the Boeing B-9 (this is a Y1B-9A test aircraft) was one of the most modern aircraft of 1931, with retractable landing gear and stressed-skin construction.

were traditional fabric and wire biplanes, although the Army did buy a few Boeing P-26 monoplanes which were again of traditional wire-braced construction, with fixed gear and open cockpits. In the sharpest contrast, American bombers led the world in the 1930s – with the possible exception of the progressively improved but basically slow Soviet ANT-6 already described, the service designation of which was TB-3. Even the TB-3 had traditional Junkers corrugated-skin construction, open cockpits and fixed gear, whereas by 1932 the Glenn L. Martin company was flying the prototype Model 123 with smooth stressed skin, fully cowled radial engines, enclosed cockpits, an internal bombload in a bay closed by powered doors, retractable land-

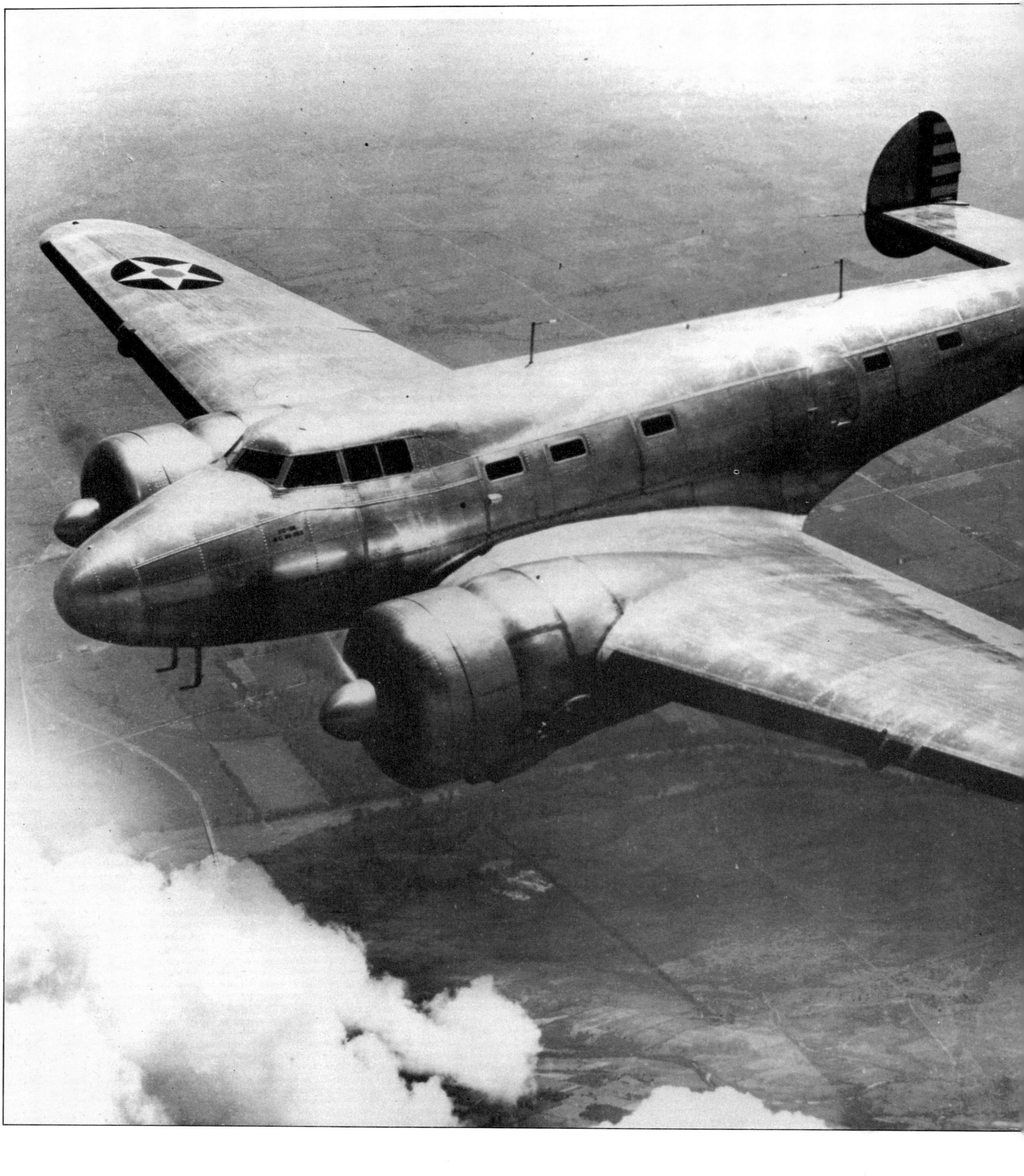

ing gear and, in a slightly modified version, a rotatable glazed gun turret above the nose. Known as The Martin Bomber, it created a sensation, not least because its speed of 207 mph (333 km/h) was faster than most 1932 fighters. Large numbers were built for the Army, and even though no exports were permitted until 1936 Martin then sold 189 more to six foreign countries.

By 1935 the Martin no longer looked like something from another planet, but even so nobody was quite ready for the Boeing B-17, the prototype of which made its maiden flight on 28 July of that year. The Army needed a new bomber chiefly to destroy any aggressive foreign fleet (no other bomber target could be imagined; the 8th Air Force of seven years later was certainly not envisaged). All that was requested was 'a multi-engine bomber', and this had in the past always been interpreted as meaning two engines. Imagine the surprise when Boeing came up with its Model 299 with four engines! This gleaming monster had its four engines not so much for speed, though it was certainly the fastest bomber in the world; they were needed for more height, because by this time there was a belief that bomber losses – from either fighters or AA guns – would be sharply reduced by flying at great heights, which then meant 25,000–30,000 ft (7.6–9.1 km). The USA led the world in turbosuperchargers, which unlike the common species geared up from the engine itself were driven by turbines spun at high speed by the white-hot exhaust gas. With turbocharged engines the mighty new Boeing promised to fly higher than all the opposition, and with the help of the Norden, a highly secret new bombsight then being designed, accurate bombing was still expected.

Sadly the Model 299 crashed on 30 October when a takeoff was attempted with the controls locked. But by this time its colossal potential was obvious and improved versions with 1,000-hp Cyclone engines were ordered for the Army. Because the otherwise streamlined fuselage was covered with gun blisters and turrets a newspaper said it was 'a Flying Fortress', and the name stuck. By 1938 improved models of B-17 were in production with turbocharged engines and revised armament, and with a speed at 25,000 ft (7.6 km) as high as 320 mph (515 km/h). At the very end of the decade, on 29 December 1939, Consolidated flew the Model 32, the prototype XB-24 Liberator, noted for its extremely long and slender wing to give high cruising efficiency for long range. Even compared with the B-17 it was an outstandingly advanced and complicated machine, with masses of electrical and hydraulic systems, and one of the new nosewheel type or 'tricycle' landing gears pioneered in the USA which offered great advantages in pilot view, ease of control and other factors over the traditional tailwheel variety.

Distinguished by its circular fuselage with small windows, the Lockheed XC-35 had turbocharged Wasp engines and set an entirely new level of performance.

Yet another form of new technology introduced in the 1930s was pressurization. One of the physiological problems of high-altitude flight is that the air pressure falls and the air gets intensely cold. Heating the cockpit or cabin of an aircraft is straightforward, but sealing it to make it airtight so that the internal pressure can be kept nearer to that at sea level is very difficult. It was easier to devise pressurized and heated flying suits, though these were very heavy and uncomfortable and had rigid helmets reminiscent of those used by deep-sea divers of that period. Such clothing was worn by S/L Swain and F/Lt Adam of the RAF when they set two new altitude records in the Bristol 138a monoplane in 1937 and 1938, only to be topped by Italian Col Pezzi in the biplane Ca 161bis at 56,046 ft (17,083 m) on 22 October 1938. But by this time several companies were already experienced with pressurized cabins.

The first to fly was a capsule made of Elektron magnesium alloy inserted into a big single-engined Junkers Ju 49 and flown on 2 October 1931. Junkers also flew the first specially designed pressurized aircraft, the first EF 61 on 19 December 1936. This was an impressive twin-engined giant with a span of 88 ft 7 in (27.0 m). Next came the Soviet Gribovskii G-14 sailplane flown in January 1937 with Shchyerbakov's first pressure capsule to fly, after which came various I-15 and derived fighters. On 9 May 1937 Lockheed flew the beautifully engineered XC-35 for the US Army, derived from the Electra but with a completely pressurized fuselage and turbocharged Wasp engines giving a speed of 240 mph (386 km/h) at 30,000 ft (9.1 km). Far more impressive than all its predecessors, the Boeing 307 Stratoliner made its maiden flight on 31 December 1938. Using essentially the wing, engines, landing gear and tail of the B-17, the Stratoliner had a giant streamlined fuselage, and though it was only pressurized to a differential (compared with the thin outside air) of $2\frac{1}{2}$ lb/sq in, compared with 10 lb/sq in for the XC-35, it was a monster fuselage seating 33 passengers in great comfort. Ordered by TWA and PanAm, it showed that the future decade need not be entirely a matter of progress only in killing machines.

The Consolidated-Vultee PB4Y-1 Liberator was the US Navy ocean-patrol and anti-submarine version of the B-24 bomber. This example has its radar extended (the B-24 had a gun turret in this location). It appears to be over England and so may be from Fleet Air Wing 7 at Dunkeswell with whom the author had the privilege of flying as passenger.

6. THE WAR MACHINES

AS THE TURBULENT, CRISIS-RIDDEN 1930s drew to a close, air forces and airlines all over the world were changing over as fast as they could from aerodynamically 'dirty' fabric-covered biplanes to stressed-skin monoplanes. It might have been expected that the two great European Fascist dictatorships, Germany and Italy, would have been right in the forefront of this process, in view of their proven technical leadership and limitless funding for war industries. In fact, Italy was strangely backward. Its aircraft continued to be a mix of steel tubes, wood and fabric, in amongst the occasional bit of light alloy, while its fighters were also relatively low-powered and deficient in speed and firepower. Here more than in any other country the view of the generals, which to a large degree reflected the opinion of the squadron pilots, was that fighters should be agile biplanes with open cockpits and two machine guns.

Japan, which from the early 1920s had carefully studied developments in other countries and at first had employed experienced foreign designers, was so little studied by Western nations that it remained an enigma. No attempt was made to discover the facts, and instead it was assumed that all the Japanese could do was make inferior copies of foreign aircraft. In fact like Italy, both the Imperial Army and Navy continued to believe in the ascendancy of manoeuvrability over speed and firepower for fighters, but it had decided by the mid-1930s that the future lay with the monoplane, and its warplanes in action over China in 1938 were as advanced as any in the world. This was especially the case with carrier-based machines, where such equipment as the Aichi D3A dive bomber and Nakajima B5N torpedo bomber were technically equal to the best in the US Navy, with stressed-skin construction, advanced aerodynamics, powerful flaps, modern engine installations, constant-speed propellers and streamlined cockpit canopies. The equivalent Royal Navy machine was the Fairey Swordfish, a fabric-covered biplane with fixed gear, open cockpits, ring-cowled engine driving a propeller of bent metal plate and so slow it needed neither flaps nor airbrakes.

Yet, paradoxically, the greatest of all wars that broke out when Hitler's Germany invaded Poland at dawn on 1 September 1939 was to show that many lumbering old types of aircraft, notably including the

Right: Screaming downhill at 180 knots the Fairey Swordfish looks archaic, but this workhorse struck hundreds of blows against many enemies in 1939–45.

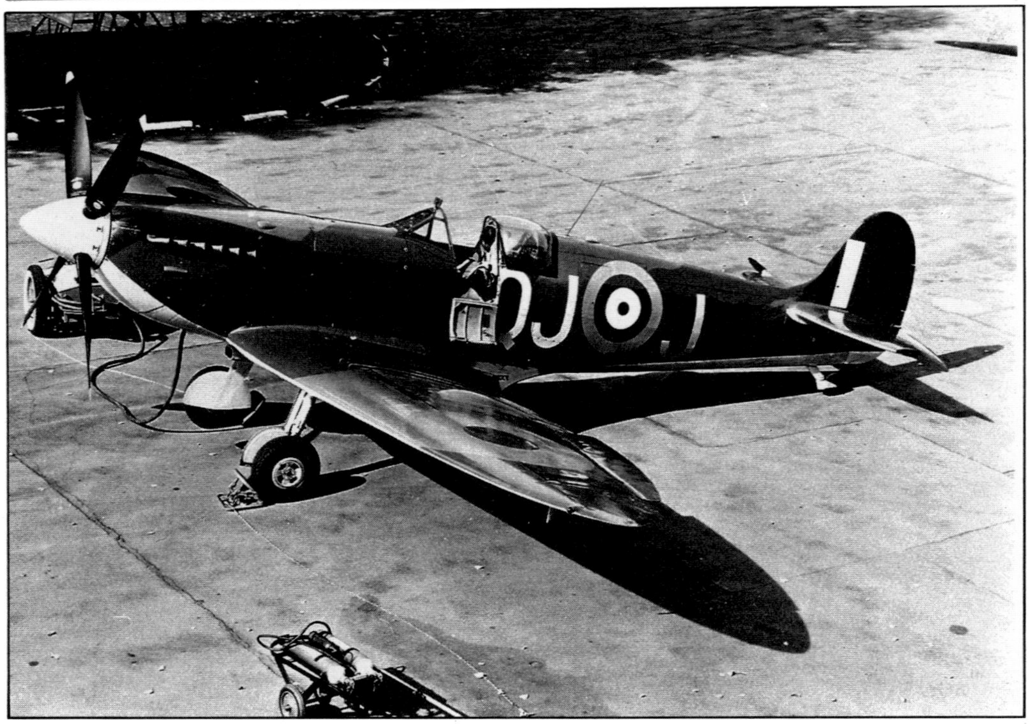

Left: This preserved Spitfire (of the RAF Battle of Britain Flight) was built as a Mk VB but has been made to look like a Mk IX, despite the pre-1942 markings! Key to the performance of all early Spitfires was the superb Rolls-Royce Merlin engine (below) which from 27 litres of cylinder capacity got as much power as enemy engines of over 40 litres.

Swordfish, were to prove more effective than their supposed replacements, and this was certainly the case with light tactical machines for close-support over the battlefield. Conversely, some of the modern warplanes that equipped large portions of Hitler's superbly trained Luftwaffe – the most feared armed force in the world in 1939 – were to prove so vulnerable to efficiently organised fighter defences that in 1940 the mighty Luftwaffe was actually to suffer a defeat at the hands of the relatively puny Royal Air Force, which had a profound effect on human history. Among the much-feared types that were to prove broken reeds were the Do 17, He 111, Bf 110 and, above all, Ju 87.

Yet at the start of World War II the Ju 87, popularly called the *Stuka* from the German word for a dive-bomber, simply blasted a path for Hitler's armies in what even then was planned in meticulous detail as the *Blitzkrieg* (lightning war). Hitler's generals had no intention of letting this war degenerate into the bloody stalemate of trench warfare. The *Panzers* (armoured land forces) and Luftwaffe were to work in partnership to roll back all Hitler's enemies almost at the best speed a tank could drive, taking prisoner all who were not eliminated. It was typical of the *Blitzkrieg* concept that the first devastating 1,102-lb (500-kg) bombs delivered by Ju 87 Stukas shouod have crashed on to the Dirschau bridge at 4.34 am on 1 September 1939, 11 minutes *before* Hitler declared war. Thereafter the *Stukas*, with about 300 being operational in the Polish campaign, demolished every target against which they were sent. Airfields, railway stations, major warships and fortifications were blasted apart with deadly accuracy, and on one occasion bombs were rained down on Polish troops a mere 100 ft (30 m) in front of the advancing German troops. A bonus was that, as the *Stuka* legend spread, the siren sound of the diving bombers struck terror even into seasoned soldiers, let alone civilians.

With total air supremacy, the Luftwaffe aircraft roamed at will across Western Poland, while of the few Polish aircraft that survived attacks on their airfields and got into the air, an estimated 35 per cent were shot down by their own anti-aircraft fire. A major contribution was made by the level bombers, the Do 17 and He 111 being backed up in this campaign by the older Ju 86 with two 600-hp diesel engines. Ju 52/3ms were the standard transports, shuttling in everything needed at the front; it is a wry comment on the Polish defences to note that of some 552 of these vulnerable aircraft engaged, only 12 were lost in action over Poland but 32 were lost from accidents during the same month of September 1939. At the final capitulation on 27 September, with Warsaw itself in ruins, Gen Loehr, commander of Luftflotte (Air Fleet) 4, said "The Luftwaffe operated for the first time in world history as an independent arm. It opened up new aspects of a strategy which in its principles had remained unaltered throughout the course of history".

This smashing success for the new *Blitzkrieg* concept and weapons had a profound effect on observers everywhere, especially in Germany. Luftwaffe Gen Albert Kesselring wrote "We learned many lessons . . . and prepared for a second, more strenuous and decisive clash of arms". In fact Hitler's efficient war machine almost had the world at its feet. It had shown, most notably by its use of airpower, that it could achieve in three weeks what had previously taken years or had not been possible at all. Had there been proper long-term planning, the Luftwaffe might have been kept so far ahead in quantity and quality of all opponents as to be able to pick them off one by one. But Nazi long-term planning was faulty, overconfident and crippled by personal whims, petty jealousies and intrigue, so that important programmes were delayed or halted whilst others were given absolute priority. But it takes time for long-term failures to become evident, and meanwhile the war could hardly have gone better.

On 9 April Hitler's aircraft and ships went north and took over Denmark and Norway. Britain tried to assist the Norwegians fight a rearguard action in northern Norway, but seaborne landings at three coastal towns between 15 and 19 April were eventually repulsed by the German invaders, and the British forces were lucky to get most of their number off in ships (this was possible only because at that time the Luftwaffe bombers seldom operated at night). A handful of British fighters, obsolete Gladiator biplanes of 263 Sqn RAF, operated briefly from a frozen lake but had to give up when the lake was heavily bombed. A squadron of Hurricanes was brought in, but owing to the general collapse was evacuated aboard the carrier HMS *Glorious*, and all the aircraft and nearly all the men aboard were lost when the carrier was sunk. Hawker Aircraft urgently flew a floatplane Hurricane, but this was not ready in numbers in time to be effective over Norway, and it certainly would have been vulnerable at its moorings and second-rate as a fighter.

The ill-starred Norwegian campaign jolted the Allies (just Britain and France at this stage) out of a relatively quiet period called 'the Phoney War'. This period had been desperately needed in order to do a little more to build up the two countries' armed forces, and especially their inadequate air forces. Britain had sent 13 RAF squadrons to France in the first week of the war, but they had accomplished little during the intensely cold winter, and the 'bomber' raids on Germany had dropped nothing but ineffectual leaflets. The only bombing raids by the RAF and Luftwaffe were against the enemy's warships. Small numbers of Ju 88s and He 111s were brought down, all by RAF fighters, over Scotland and the Shetlands. Against Germany the RAF fared worse; on the second day of the war a force of Blenheims and Wellingtons, the latest and best RAF bombers, suffered a loss rate of 37 per cent, while on 18 December a tight formation of 22 Wellingtons, each with a power-driven gun turret at nose and tail, was intercepted by Messerschmitts and in a few minutes 12 were shot down and three badly damaged.

What Britain never suspected – despite having been given all the details by a well-wisher in Oslo in November 1939 – was that Germany had a chain of operating stations, and that one of them had been plotting the British raid of 18 December from 70 miles (113 km) away. Without any evidence, Britain believed that it alone had radar; and the fact that it did was purely by chance. In 1934 there was no scientific basis for air defence, other than laborious mathematical calculations about AA guns and their problems, and cumbersome sound locators which enabled the direction of the incoming noise of an enemy aircraft to be plotted. Sound locators were not accurate enough for guns to be aimed by them, and in any case as the sound from a bomber ten miles (16 km) up – even directly overhead – takes almost exactly a minute to reach the ground, in which time the target would probably have changed course, it was pretty useless. Britain even constructed gigantic concrete 'sound mirrors', which still exist near the English coast as monuments to a lack of any attempt to discover a better answer.

In 1934 there was much talk of 'death rays' and the Air Ministry in London asked a radio expert, Robert Watson Watt, whether such a ray could be devised. The answer was negative; but Watson Watt pointed out that you first had to discover where to aim the ray, and quite casually said this could be done by radio (nobody had ever asked him this question!). The result was a total revolution in warfare in the rapid development of radar. Other countries, notably the USA and Germany, were also well advanced with the idea, but Britain was first in constructing an air-defence chain of radar stations, which from 1936 gradually extended along England's east and south coasts until by 1939 there were no gaps. With skilled girl plotters it became possible to detect single aircraft or formations almost as soon as they took off on the Continent and to plot their progress accurately towards Britain. At last the defending fighters could be steered accurately towards the enemy. Previously squadrons had had to take off all round the clock and maintain standing patrols in the hope of catching sight of a hostile bomber. With radar, there was no need to take off unless the enemy was coming; but once in the air, the defending fighters could be 'vectored' by a ground controller into the ideal position for shooting the enemy down.

Yet the first aircraft actually to do this were the Messerschmitts of the Luftwaffe. Germany had developed a range of radars by 1939, some for installation in warships and others for air-defence on land. Chief among the latter were giant Freya long-range warning sets, and Würzburg precision-guidance radars with large steerable dish aerials for directing the fire of Germany's formidable *flak* (AA guns). Obviously, radar works by day or night, but night operations by military aircraft had always been limited since 1918. The many wars of the 1930s had seen hardly any night flying, and at the start of World War II there were no major plans to do any. But the shattering losses suffered by RAF bombers over Germany, without even crossing the coastline in any significant way, was a rude shock. Such casualties could not be sustained, and so the useless but gruelling leaflet raids during the winter 1939–40 all took place by night.

But it was gradually evident to the RAF, and to some degree to other air forces, that night flying was still a tricky and largely unacquired art. Peacetime training had been inadequate in amount and in scope, so that – in the words of a Bomber Command squadron leader in 1940 – "we were lost as soon as we left the aerodrome". Not untypical was the bombing raid on the naval base on the island of Sylt on the night of 19 March 1940. Of the 50 RAF crews taking part 41 reported they had bombed the correct target, but subsequent air reconnaissance showed that not one bomb had fallen on it.

During the first four months of 1940 more than 2,000 aircraft were delivered to units of the RAF and Armée de l'Air, more than 600 having been supplied from the USA (mainly to France). The American types especially included the North Ameri-

Left: An RAF recruiting centre photographed in 1940. The sergeant air-gunner has already seen active operations, judging by the bars on his uniform.

Below: The original Sunderland I introduced the RAF to a new type of aircraft, more powerful than anything previously dreamed of. Despite its bulk it was faster than many fighters still in RAF squadrons.

can NA-16 trainer, already widely used as the BC-1, BC-9 and AT-6 in the USA and later to be the most important trainer of all time with almost 20,000 in many versions (the British Commonwealth name was Harvard). US combat types included the speedy Douglas DB-7 and Martin 167 tactical bombers and the quite effective Curtiss Hawk 75A fighter, with a constant-speed propeller and six machine guns, which was much better than the Morane-Saulnier M.S.406 which was the main French fighter. The Dewoitine D.520 was rather better, but few were available when Hitler ended the Phoney War by striking in the west on 10 May 1940.

To say that this was the ultimate triumph of *Blitzkrieg* is to understate the case. No radical new weapons or techniques were used, except for the bold despatch of a mere 70 airborne troops in seven small DFS 230B-1 gliders which landed right on top of the biggest fortress in Belgium, Eben Emael at the junction of the Maas with the Albert Canal. This puny force crippled the gigantic fortress and finally accepted the surrender of its demoralized complement, at the cost of only five casualties. Other airborne forces came to over 40 key points in the Low Countries by Ju 52/3m, He 59 and Ju 52 seaplanes, DFS 230 gliders and by parachute, to seize and hold many vital objectives. Defending air forces were destroyed on the ground, their airfields being left as undamaged as possible for subsequent use by the Luftwaffe.

France was too big for speedy envelopment, but the crucial superiority of the Messerschmitt Bf 109, by now mainly in its more powerful E (Emil) version with the DB 601 engine, sufficed to secure and hold complete air supremacy. This caused colossal losses to the Allied aircraft and maintained ideal unworried operating conditions for the *Stukagruppen* which caused such fear and paralysis among the defending armies that Guderian's *Panzers* had to keep revising their schedules because of

Right: Over 4,000 PBY Catalinas were built in many versions, making this the most numerous water-based aircraft of all time. This is a US Navy PBY-5A amphibian, with radar.

Below: Nearest the camera, the Lockheed F-5B was one of the unarmed photo-reconnaissance versions of the famed P-38 Lightning long-range fighter (seen in the rear). Powered by two turbocharged Allison engines, this unusual aircraft was probably the quietest fighter ever put into service.

267183
267332

Above: Seen at Brooklands just before its first flight on 6 November 1935, the Hawker High-Speed Monoplane (F.36/34) was the forerunner of the Hurricane.

their rapid progress. It is a reflection on the situation to note that in fact, the Bf 109s, Bf 110s, Do 17s, He 111s, Ju 87s and Ju 88s of the Luftwaffe were far outnumbered by the more than 40 types of combat aircraft flung into the fray by France, Britain, Belgium and the Netherlands, yet the Allies were steadily destroyed and put to flight.

There was intense pressure on Air Marshal Hugh Dowding, C-in-C of RAF Fighter Command, to send more and more squadrons to France. He had the political courage to refuse, seeing that the defence of France was already hopeless and that further attrition would merely imperil that later Battle of Britain that was bound to follow. But it was clearly right to send as many RAF squadrons as possible to try to protect the beaches of Dunquerque where, from 26 May (a mere two weeks after the start of the German attack), the remnants of the British army, with a few Allied troops, were evacuated by an armada of small ships. The whole might of the Luftwaffe's I, II, IV and VIII Fliegerkorps was thrown into frustrating the evacuation, but this time the Messerschmitts met the Spitfire for the first time. Suddenly they were up against really tough opposition, and a combat could go either way. Against all the odds, the Luftwaffe was prevented from stopping the giant evacuation, which brought off over 335,000 men by 4 June.

While the victorious Luftwaffe rested and refitted in its new bases throughout the northern half of France (the southern half remaining in the hands of a puppet government), the beleaguered RAF enjoyed a further respite of six weeks before *Adler Tag* (eagle day), originally set for 13 August 1940. This was to be the start of the final campaign in the west, the elimination of the RAF and the conquest of Britain.

Meanwhile, since 30 November 1939 the so-called Winter War had been in progress in Finland. The mighty Soviet Union, which had a non-aggression pact with Hitler and had occupied the eastern half of Poland in September 1939, had picked a quarrel with Finland and invaded on that date. The Soviets attracted international criticism because of the apparent unjustness of their campaign, and this was fuelled by the courageous and dogged defence by the tiny Finnish armed forces. Deploying an initial total of 116 mainly obsolete aircraft, the Finns fought the vastly superior Soviet air forces almost to a standstill in the severe winter, but could not prevent Helsinki, Abŏ and other cities from being heavily bombed. The Western nations hastened to supply a handful of Fokker D.21s, Curtiss Hawk 75s, Gladiators, Hurricanes and Brewster B-239 Buffaloes, but as the winter receded Soviet air commitments were stepped up to over 2,200. These comprised chiefly Polikarpov I-15 biplane fighters, Polikarpov I-16 monoplane fighters, Tupolev SB-2 medium bombers, Ilyushin DB-3B bombers and a few of the giant Tupolev TB-3 four-engined heavies. The Finns were forced to seek surrender terms on 13 March 1940.

By the summer of 1940 more than 4,000 aircraft of some 160 types had by various ways been painted with British markings since the start of the war. Some were regular deliveries of such production types as the Whitley and Wellington heavy bombers, the efficient but poorly protected Hampden medium bomber, the light Blenheim and Battle attack bombers (both of which had been proved unacceptably vulnerable during the fighting over France), and the Hurricane and Spitfire fighters. But most were refugees from all over Europe, and hordes of former civil aircraft from single-seaters to airliners, all of which either had been registered in Britain or sought refuge there. A few were useful over a long period, but the vast majority soon ran out of hours and, cut off from foreign spare parts, gradually rotted on the growing sprawl of British airfields. The latter were constructed under high pressure until their circuits overlapped in many parts of England, and the total by late 1944 exceeded 1,500.

While workers toiled in Britain to build up the RAF, and Fighter Command in particular, the aircraft industry competed with all others for scarce labour and materials. From the outbreak of war, the undersea menace of the German U-boat had grown swiftly, and in the first winter a further menace was the very effective German magnetic mine, laid not only by ships but also by such aicraft as the He 115 floatplane seaplane, which flew almost with impunity along the Thames and up the east coast at low level, within a few metres of the defending radar towers – but by night. The night defence of Britain was almost non-existent in early 1940, and most other countries were in the same state. But following the bombing of Rotterdam during the initial German assault to the West the new British Prime Minister, Churchill, gave the RAF permission at last to bomb Germany. The result was the appointment of Gen Josef Kammhuber to form a night defence force. He gradually built up a system of *Himmelbett* (four-poster bed) blocks of airspace in each of which was a Freya long-range surveillance radar and a Würzburg precision radar with a steerable dish aerial, the idea being that the two radars in combination would be able to guide a night fighter astern of a selected

Right: Suddenly in 1938 the RAF was allowed to buy hundreds of new aircraft and US industry was harnessed to help. Here Hudsons for Coastal Command are hustled down the Lockheed assembly line.

Right below: Bell P-39 Airacobras ready for collection from Niagara Falls/Buffalo in 1942.

24387
24388
HOIST
173

RAF bomber. But at first success proved very elusive, and the only significant result was attrition of the force of Bf 110 heavy twin-engined fighters.

The Bf 110, first flown in 1936, was planned as the spearhead of the Luftwaffe, the *crème de la crème* among Goering's handsome élite. As fast as the best single-engined fighters, it had a heavy armament of two cannon and four machine guns firing ahead, with another machine gun providing rear defence, and its large fuel capacity enabled it to fly long missions quite beyond the capability of smaller machines. In Poland it swept all before it, and in the Norwegian campaign it proved absolutely invaluable. Indeed one local *staffel* commander, running out of fuel over Oslo's airport at Fornebu, boldly decided to lead his Bf 110s of 1/ZG 76 down to capture the

Right: Deck parties pull the chocks from under the wheels of an A6M2 (Zero) aboard the carrier Hiryu *at the start of the Pearl Harbor raid.*

airport by themselves, without waiting for the big force of delayed paratroops! But in May and June 1940 the Bf 110s often met fairly modern single-engined fighters and found that they were too cumbersome to fight them.

Thus the stage was set for the biggest and most decisive air battle of all history up to that time, the Battle of Britain. The conflict started with local battles over the Channel as *Stukas* dive-bombed British ships. Since early June Fighter Command had managed to increase its front-line strength from a mere 386 Hurricanes and Spitfires up to almost 600; but against them were ranged over 2,600 modern aircraft, crewed by men flushed with success, well trained and in general possessed of vastly greater combat experience (many had fought in Spain). The conflict looked very one-sided, and as it opened by deliberate attacks on the vital British radar stations round the coast there seemed certain to be only one eventual outcome.

Right: Flight crews of the Imperial Japanese Navy get ready for a mission with G4M2e bombers each carrying an MXY7 suicide piloted missile.

At the outset even the RAF's tactics were faulty, the unwieldy squadrons in V formation being completely outmanoeuvred by the fluid Messerschmitts in *Rotte* (loose pair) formation. Gradually Fighter Command learned to operate in the same flexible arrangements as their opponents, to keep looking for 'the Hun in the Sun' as their fathers had done in 1916, and to assign 'weavers' to keep manoeuvring and searching the sky astern. They also learned to fly their Hurricanes, which equipped 33 squadrons, and the Spitfires, which equipped 15½, to the absolute limit, using maximum power and maximum manoeuvrability. Moreover, the performance of these aircraft, marginally adequate for the task, had recently been dramatically improved. The ancient wooden propellers were progressively being replaced by new variable-pitch or even constant-speed three-blade metal propellers made by Rotol or de Havilland, giving much better takeoff and climb and higher top speed. Even more added performance came from the highly secret introduction of 100-octane fuel, not manufactured in Britain at that time and available only from Baytown, New Jersey, and Aruba in the Caribbean.

Left: With over 10,000 built, the A6M 'Zero' was the main Japanese warplane of World War II. This is an A6M5 captured by the US Army Air Force.

This greatly increased the power that could be wrung from the Rolls-Royce Merlin engines, if anything giving the RAF fighters an edge over the Bf 109Es. Yet the latter continued to have two advantages that could not immediately be countered: its engine had direct fuel injection instead of a float-type carburettor, so it kept running if the pilot thrust his stick forward into a steep dive, and it had two 20 mm cannon in its armament which could destroy a fighter with one shell at ranges beyond the limit for the eight small machine guns carried by the British machines.

Despite this, the 109s for the first time found themselves at least equalled, if not outfought. There were increasingly grave errors in the Luftwaffe's conduct of the battle, especially when Goering or Hitler issued direct orders, so that although the RAF squadrons had no respite, at least their remnants could be rotated to rear areas for rest and re-equipment. Attacks on airfields were never continued over a period to render them unusable, and with one exception the radars were never 'off the air' for more than a few hours at a time. Most of the major factories making fighters and their parts, notably the vital Merlin factory at Derby, were left in full production, working round the clock. Moreover, the Luftwaffe suddenly learned that its chief aircraft types were vulnerable, apart from the 109. The fast bombers, each with three hand-aimed machine guns, were downed in droves. Hasty lash-up schemes included an arrangement of four separate machine guns in the Ju 88, all to be aimed and fired by one man! The long-range Bf 110, intended to protect the bombers, was so easy to shoot down that it needed protection itself, and was soon withdrawn from western Europe. As for the previously feared Ju 87 *Stuka*, this was shot down with such ease that from 18 August it ceased to come to Britain. Luftflotte 5, based in Norway, made what it thought would be an easy mass attack on north-east England on 15 August, only to find that the RAF fighters had not all been moved to the south; it lost 16 bombers and seven Bf 110s and did not appear again. In southern England the short range of the Bf 109, and the order that the bombers should be escorted closely, negated the German type's combat advantages, so that the outcome hung in the balance for weeks. By the start of September Fighter Command was in a crisis, mainly because its experienced pilots had been killed and the replacements were easy meat for the skilled Germans. But the tide turned on 4 September when Hitler switched the attack to London; while pressure on the RAF fighters eased, the Luftwaffe losses increased.

At night the RAF had steadily bombed the invasion barges moored around the European coast, and on 17 September, two days after one of the biggest daytime battles of all, Hitler ordered the invasion plan indefinitely postponed. He set his sights instead on the Soviet Union, leaving England to be attacked at night, with daytime raids restricted to Bf 109Es carrying 551-lb (250-kg) bombs. In fact the US Navy had slung bombs under its carrier-based fighters since 1929, but in Europe it was novel. High-speed fighter-bombers were to raid the coastal towns of England until 1944, though with increasing losses. Meanwhile the RAF put more powerful Merlins into its fighters and set about revising the armament, adding first two or four 20-mm cannon, then 250-lb (113-kg) bombs and by 1942 batteries of rockets with 60-lb (27-kg) explosive warheads, which were highly effective against tanks, trains, buildings and small ships. Throughout 1941 fighter 'sweeps' carried the war to the enemy, though the results hardly justified the losses – especially after the Focke-Wulf Fw 190 was encountered in numbers. Powered by a large BMW 801 radial engine, this extremely compact fighter-bomber had flown at Bremen airport in June 1939 but was totally unknown to British intelligence. It outfought the latest

Below: USS Yorktown *lists to port at noon on 4 June 1942 during the Battle of Midway. She was later abandoned, but US Navy SBDs sank four Japanese carriers.*

Above: F4U-1 Corsairs of US Navy squadron VF-17 became operational at New Georgia in 1943. The bent-wing fighter was one of the best warplanes.

Spitfires, but by sheer chance Rolls-Royce had already produced a completely revised Merlin with two superchargers in series which, fitted to the new Spitfire IX, in 1942 at last restored a measure of parity.

Thwarted by day, the Luftwaffe raided Britain by night, and at first the defences were ineffectual. In 1938 the first experiments had been made in trying to develop a small radar that could be fitted into a fighter, and a handful of radar-equipped Blenheims were flying when war began. But the first effective night fighter was the Bristol Beaufighter, a robust and powerful machine with two 1,650-hp Hercules sleeve-valve engines and carrying not only radar but a backseat crew-man to operate it, fuel for an all-night patrol and the unprecedented armament of four cannon and six machine guns. Gradually the temperamental radar was mastered, and by May 1941 Beaufighters were causing significant attrition to the Luftwaffe. On 31 May the heavy assaults ceased, because the Luftwaffe was needed elsewhere. The first thrust was south-east through the Balkans, where Hitler's incompetent ally Mussolini had been fought to a standstill by the gallant Greeks. In North Africa vast numbers of Italians had been taken prisoner by British and Commonwealth forces, and Cyrenaica occupied as far west as El Agheila. The Italian fighters, the Fiat C.R.42 biplane and G.50 monoplane, and the more streamlined Macchi C.200, had been outfought even by the rather second-rate collection of British machines that could be spared for this theatre, and only the three-engined S.M.79 had achieved any success in its role as a bomber and torpedo carrier over the Mediterranean. Hitler sent an Afrika Korps, backed up by quite modest Luftwaffe units which included two *gruppen* of Bf 109s – I/JG27 and 7/JG26 – and these turned the tables and quickly achieved smashing success. A far larger force, Fliegerkorps X, was sent to Sicily on 12 January 1941 to bomb Malta into submission and stop British ships from using the Mediterranean. On 6 April came the mighty thrust through Yugoslavia and Greece, culminating in the air-launched invasion of Crete, which was finally evacuated by Commonwealth troops on 31 May after heavy casualties on both sides. Just what the *Stukas* could do when the RAF was absent is shown by the Royal Navy's losses in the final week of May, off Crete: three cruisers and five destroyers sunk and three battleships and two cruisers badly damaged.

Suddenly, on 22 June 1941, Hitler launched Operation Barbarossa. The biggest army and air force the world had yet seen struck east, into the totally unprepared Soviet Union. It seems incredible that an advance was possible along a front over 1,200 miles (2000 km) long, and that it should have been calculated that armies of 4,000,000 and an air force larger than the Luftwaffe should be destroyed in six weeks; but for the first few weeks everything went according to plan. The *Panzers* moved ahead in a series of giant pincer movements, each swallowing up to 250,000 men and vast amounts of weapons, while the initial assault from the air destroyed 1,811 Soviet aircraft for the loss of just 32; over 20,000 Soviet aircraft were destroyed by October. But it was not all over in six weeks; winter set in, the whole of Soviet industry was evacuated thousands of kilometres east to beyond the Urals, and German losses had been on an unprecedented scale. The vast German war machine was suddenly overstretched.

By chance the Soviet VVS (Air Force) was caught at the wrong moment. Its thousands of I-15 and I-16 fighters and SB-2 bombers were obsolete and could hardly blunt the onslaught of the Luftwaffe. The next-generation fighters, the MiG-1 and -3, Yak-1 and wooden LaGG-1 and -3, were not only undergunned but very tricky to fly; they were dangerous to land on the poor Soviet fields and suffered very heavy casualties quite apart from those inflicted by the enemy. Meanwhile the Luftwaffe aces, many already highly experienced, shot down the Russians in droves, so that eventually more than 100 had personal scores exceeding 100; 35 of them surpassed 200, and the top scorers were Majors Gerhard Barkhorn (301) and Erich 'Bubi' Hartmann (352). Throughout, the main Luftwaffe fighters continued to be improved versions of the Bf 109 and Fw 190, the later 109s being the G (Gustav) model with the more powerful DB 605 engine. By the end of the war some 33,000 109s had been made, mainly because nothing arrived in time to replace it. It certainly swept all before it in the Soviet Union, so that few Russian pilots survived long enough to learn their trade.

Among Soviet bombers the DB-3B had been improved into the Il-4, of which 5,256 were produced. It flew many long missions with bombs or torpedoes, but for much of the war the front line was too far east for targets in Germany to be reached. The obsolete TB-3 was succeeded by the TB-7 (Pe-8), but this big machine was made only in trivial numbers. At the opposite end of the scale, the Po-2 biplane, a simple 100-hp trainer first flown in 1927, was used in vast numbers as a front-line combat machine with bombs, guns and other weapons; it even had psy-war loudspeakers and ceaselessly harrassed the Germans in their bunkers, tanks and buildings (it was said "The Po-2 comes and looks over the

Right: Hauptmann Franz von Werra as CO of I/JG53 in the Soviet Union after his escape from British captivity. He dived into the sea on 25 October 1941 and drowned.

Below: One of the most important aircraft types in the Pacific was the F6F Hellcat. Here F6Fs share the deck with the other war-winner the SBD Dauntless.

Right: The North American B-25 Mitchell was one of the chief Allied tactical bombers, serving on all fronts. These are Mitchell IIs of RAF No 180 Sqn, West Raynham.

Right below: Over 87 ft (27 m) long, the Short Stirling I was expected to be 'the bomber to win the war'; in fact it was inferior to the Lancaster and Halifax.

W
NULLI SECUNDUS

E

window sills to see if we are inside"). By 1943 the Germans also had scoured Europe for old biplane trainers which were organised into light attack units for the front line and to operate against partisans in the rear.

Most important of the attack aircraft on the Soviet side were two real war-winners, the Petlyakov Pe-2 and Ilyushin Il-2. The Pe-2 was a very fast twin-engined bomber; when RAF No 151 Wing took its Hurricanes to the Murmansk sector in 1941 they were amazed to find that they could hardly keep up with the Pe-2s they were detailed to escort. Altogether 11,427 of these tough and useful three-seaters were delivered, including many fighter, reconnaissance and specially armoured models. The single-engined Il-2 was the arch-type of the *Shturmovik* (armoured attacker), and its crew of one or two sat surrounded by heavy armour which also protected the engine and other vital items. Thus it was much harder to shoot down than the British Fairey Battle which it otherwise resembled in size and shape. Moreover, unlike the Battle, it carried heavy cannon and rockets, as well as bombs, and throughout the war took steady toll of the German *Panzers*, including even the massively armoured Tiger. No fewer than 36,163 of these tough birds were made, more than for any other single type of aeroplane.

On a front as vast as this, air transport played a central role. Back in 1940 Hitler had realized that giant gliders would be needed to bring heavy guns and *Panzers* to Britain along with the airborne troops. Messerschmitt produced the Me 321 Gigant, a colossal glider made of steel tubing, with ply or fabric covering. Carrying a load of 22 tonnes, or a complete company of infantry, it had a flat level floor accessed by the entire nose opening into left and right halves. Great difficulty was experienced in towing it, even with booster rockets added under the wings at takeoff. One answer was the

Above: Two Grumman F6F-5 Hellcats flying near the Brooklyn Bridge just after the war, with a reserve unit at Floyd Bennett Field.

Right: When the Grumman TBF-1 Avenger torpedo bomber first went into action at the Battle of Midway in June 1942 the result was disastrous.

troikaschlepp: three Bf 110s pulling the monster via three steel cables. This caused hair-raising incidents and a few crashes; a better answer was the unique He 111Z, in which two He 111 bombers were joined together via a common centre section with a fifth engine. But even this was not a good idea as a regular trucking system, and this is what was needed to sustain the surrounded German VIth Army at Stalingrad in November 1942. Frantic efforts at air supply, involving not only every available Ju 52/3m but also the giant Ju 290, He 111

and He 177 bombers and the Gigant gliders, proved inadequate, and on 2 February 1943 the 91,000 survivors (including 24 generals) of 220,000 finally surrendered. At the same time the last remnants of the vastly augmented but defeated Afrika Korps were fighting a rearguard action in Tunisia, finally withdrawing to Sicily on 12 May 1943. The Allies followed to Sicily on 10 July, and to Italy itself on 3 September. Italy capitulated on 8 September, and subsequently dogged resistance was put up in Italy by German forces commanded by Luftwaffe Gen Albert Kesselring. New weapons used in summer 1943 included the German Hs 293 radio controlled rocket missile and FX.1400 radio-guided bomb, both used by Luftwaffe Do 217K-2s and other bombers against Allied ships. FX, or Fritz X, had a devastating punch, and a single hit on the Italian battleship *Roma* – a modern ship of 46,200 tons – sent her to the bottom as she steamed to join the Allies on 9 September. Her sister *Italia* also took a hit but managed to limp to Malta.

Below: Handley-Page Halifax heavy bombers were responsible for countless nocturnal raids over the Third Reich in addition to their work on Maritime reconnaissance, paratrooping, glider-towing, anti-submarine work and general duties.

In the crucial Battle of the Atlantic the U-boats became ever more deadly, sinking astronomic tonnages of Allied vessels and threatening to bring the whole British war effort to a standstill. RAF Coastal Command expanded enormously, adding white-painted Whitley and Wellington bombers fitted with ASV (anti-surface vessel) radar, new versions of the Sunderland and Catalina flying boat, and heavily armed Beaufighters to sweep away Luftwaffe prowlers from Iceland to Gibraltar. Gradually the Fw 200C Condor long-range bomber, which, Churchill called 'the scourge of the Atlantic', was mastered by a combination of Beaufighters from Britain and such better naval fighters as the Sea Hurricane, Grumman Wildcat (the US Navy F4F) and, from late 1942, the Seafire, the naval version of the Spitfire. During the dark days of 1941 Hurricanes had even been carried on CAM (catapult-armed merchantman) ships, blasted off by rockets if a Condor should appear, thereafter to do a ditching in the freezing ocean so that, with luck, the pilot might be rescued. The U-boat campaign came to a mighty crescendo in May 1943, with hundreds of Type VIIC boats operating as wolf packs and, fitted with special ECM and receivers tuned to RAF radars, fighting it out on the surface with deadly batteries of flak. The ultra-long-range Liberators closed the last mid-Atlantic gap where previously air power could not reach, and by the end of May the U-boats were at last defeated. Survivors slunk home, never again to prove a major threat. But convoys to the Soviet northern ports continued to be heavily attacked, notably by the Ju 88s of KG30, though never so heavily as luckless convoy PQ17 which was decimated in July 1942. Among the war material sent to the Soviet Union were some 15,500 aircraft from the USA and over 4,000 from Britain, vast amounts of machine tools and enough aluminium (250,000 tons) for two years of aircraft production at the 1944 rate.

Right: *An 8th Air Force B-17 in flight over England.*

American war production had risen enormously in 1940–41, even though the nation was neutral and isolationist, much of the increase going to Britain under direct commercial contracts. Despite great differences in instrumentation, systems design and operating techniques, most of the American types supplied to the RAF were outstandingly successful, examples including the Douglas Boston light bomber and its Havoc night-fighter version, the Lockheed Hudson ocean recon aircraft, the Catalina flying boat with 30-hour endurance, the square-winged Grumman Wildcat

naval fighter and a succession of versions of the very complex and challenging Liberator which, because of its unrivalled long range, was used for many bomber, transport and ocean patrol tasks. Such types as the Martin Maryland and Baltimore tactical bombers and mass-produced Curtiss P-40 Tomahawk and Kittyhawk fighters were merely adequate, and served intensively in the Mediterranean. But some of the most advanced machines, including the Bell P-39 Airacobra fighter, with tricycle landing gear, car-type doors and the engine behind the pilot, the very fast Lockheed P-38 Lightning twin-engined fighter with a short cockpit/armament nacelle on the centreline and the tail carried on twin booms, and the Boeing B-17C Fortress high-flying bomber, with turbocharged engines and 320 mph (515 km/h) speed but only a modest bombload, all proved to be unacceptable to the RAF. Part of the trouble was unfamiliarity and incorrect operating procedures, and all three were used in great numbers by the US Army Air Forces.

The USA had been violently drawn into the war by the sneak attack of Japan on the US Pacific Fleet at Pearl Harbor, Hawaii, on the morning of Sunday 7 December 1941. Over the subsequent three months not only did the Japanese blow sky high the myth that their aircraft were inferior copies of Western types but they conquered the whole of eastern Asia, hammered at the gates of India, took over all the vast spread of islands today called Micronesia and Indonesia, and reached the shores of Australia – by far the largest area ever conquered by a single nation. Their total dominance in the air was possible mainly because of the indifferent opposition. In fact the two principal Japanese types of fighter, the Army Ki-43 Hayafusa (peregrine falcon) and Navy A6M Zero-sen, were very ordinary, lightly built machines, with engines of barely 1,000 hp and quite light armament (especially the Ki-43 which for most of the war had only two machine guns). Their great attribute was outstanding manoeuvrability, and, coupled with rapid and steep climb, good fuel capacity and remarkably low fuel consumption, they were well suited to the challenging task of destroying air opposition over enormous areas of sky. Indeed, the Zero at first gained a reputation of invincibility that was damaging to Allied morale. It should not have been such a shock, because detailed reports on it had been sent to Washington from China, where it first

Left: Yokosuka naval base, Japan, photographed 18 April 1942 from one of Jimmy Doolittle's token force of B-25 Mitchells on their one-way mission from USS Hornet.

Left below: First supplied to France (as the DB-7) and the RAF (as the Boston) the Douglas A-20 was one of the best twin-engined attack aircraft of the war.

appeared, from August 1940 onwards! Not until July 1942 was an example captured – it had made a forced landing on a remote island in the Aleutians – and brought to the USA. Then it was found to be far from invincible and to possess many shortcomings.

A general fault of the Japanese aircraft was light construction in the interests of manoeuvrability for fighters and range for bombers, so that they proved singularly vulnerable when intercepted. The latest Imperial Navy bomber, the Mitsubishi G4M – called 'Betty' by the Allies, who invented their own names for the largely unknown enemy machines – ought really to have been a four-engined aircraft because of the severe demands for great operating range with a heavy bombload. With just two 1,800-hp engines it had to be deficient in self-sealing tanks and armour, so that it burned so readily that Allied fighter pilots later called it 'the honourable one-shot lighter'. Back in the terrible winter of 1941–42 it was rather different. Only three days after Pearl Harbor the two British capital ships in the Far East, HMS *Prince of Wales* and HMS *Repulse*, were caught by a small group of G4Ms, with some older G3Ms, and sent to the bottom with great loss of life. On 15 February Singapore, its air defences decimated, finally surrendered to an enemy who had come not by sea but by land, where the big guns of the island could not be brought to bear. In the first week of April Aichi D3A dive bombers sank the remaining British major warships in the Indian Ocean, the cruisers *Cornwall* and *Dorsetshire* and the carrier *Hermes*.

By sheer chance the American carriers had been away from Pearl Harbor on 7 December, and their complements of F4F fighters, Douglas SBD dive bombers and TBD torpedo bombers were intact. In early May they met the Japanese fleet in the Battle of the Coral Sea, and although both sides made many mistakes the US Navy had the better of it. Admiral Yamamoto was convinced the US carriers were out of action, and he staked all on attacking an island group – Midway – so important that the US Navy would have to send the rest of its fleet, which could then confidently be destroyed. But his coded signals were deciphered, and by superhuman efforts the carriers *Yorktown*, *Hornet* and *Enterprise* were readied and reached Midway on 4 June 1942. The three-day battle was one of the decisive ones of the war. Yamamoto

Right: The Armstrong Whitworth Whitley (this is an early Mk II) was the first of the RAF's new heavy bombers ordered under the 1937 expansion scheme.

Right: Only a few Mk XVIII Mosquitoes were built, with the usual four cannon replaced by a 6-pounder gun, for use mainly against ships. Machine guns were retained.

sent his A6Ms, D3As and B5N torpedo bombers against Midway, but met strong opposition. They returned to their carriers and, while they were being refuelled and rearmed, the three US carriers struck with full force. The three great Japanese carriers *Kaga, Akagi* and *Soryu* were all destroyed. Only *Hiryu* survived, to cripple *Yorktown*, but on 6 June this carrier too was caught and sunk. It was the turning point in the Pacific. Japan could build more aircraft but she could not quickly build more carriers, nor replace 332 experienced pilots.

Japan's great offensive reached stalemate in New Guinea, and in August the Americans invaded the Solomon Islands, triggering off the bitter six-month Battle of Guadalcanal. Before Pearl Harbor Admiral Yamamoto had said "I fear we are awakening a sleeping giant'. Now the truth of his prophecy began to be self-evident, as the gigantic US industrial machine turned to making weapons. Not least of these weapons were the Vought F4U Corsair, Grumman F6F Hellcat and Grumman TBF Avenger. The first Corsair had flown back in May 1940; it was large for a single-engined fighter, with strange bent wings (the so-called inverted gull form) partly to match short landing gears with enough ground clearance for the giant propeller, driven by one of the big new Pratt & Whitney R-2800 Double Wasp engines, which started life at 2,000 hp. In October 1940 the first F4U had been the first fighter in any country to exceed 400 mph (644 km/h). It took a long time to get it right for use, and even then only the Royal Navy managed to operate it from carriers, but many consider it the best all-round fighter of the whole war. It devastated the Japanese, and also blasted them on the ground with bombs and rockets. But the chief architect of Allied air superiority in the Pacific was the F6F, which did not fly until June 1942 but very quickly got into production at such a rate that over 12,000 were made in one factory in two years. From 1943 the F6F operated from all the big carriers and most of the small escort 'flat tops' as well, destroying Japanese aircraft of all types, wherever they could be found. Its partner the TBF, was the new torpedo bomber, but the old Douglas SBD, named the Dauntless, continued in production and, emphatically not being replaced by the much more expensive new SB2C Helldiver, actually sank more Japanese ships than any other Allied weapon.

Douglas also made no fewer than 7,385 of the fast bomber series known as Bostons to the RAF but as A-20s to the US Army. In the Pacific, as on other fronts, it was partnered by the bigger but slower North

Above: Fitted with a highly supercharged Merlin engine the North American Mustang became a war-winner. This is a P-51B of the 354th Fighter Squadron.

American B-25 Mitchell, which gradually grew extra guns until some had 18 of the hard-hitting Browning 'fifty calibre' (0.5-in/12.7-mm) size as well as bombs. Back in April 1942 Jimmy Doolittle, by this time a colonel, led 16 early B-25s on a daring mission from the carrier *Hornet* to bomb Tokyo and other Japanese cities. The B-25 had never been designed to fly from a carrier, and there was barely enough run for the overloaded bombers to stagger into the air, but the B-25s bombed Japan and then went on to land wherever they could in China. A third twin-engined bomber was the Martin B-26 Marauder, powered by two Double Wasps and initially called 'the Widow Maker' because pupils found it difficult to fly. It went on to have one of the lowest loss rates of any Allied aircraft, and well over 5,000 were built.

The famed 'Doolittle raid' had been undertaken mainly as a morale-boosting publicity stunt, and the same could be said of the foolhardy daylight raid mounted by RAF Bomber Command on 17 April 1942. The MAN diesel plant at Augsburg, in southern Germany, was the target of 12 of the new Avro Lancasters, which had hardly ever been on operations before. Air Marshal 'Bertie' Harris, better known later as 'Bomber' Harris, took over Bomber Command on 22 February 1942. With Churchill behind him, he was determined to bomb Germany into submission. Almost the whole command was equipped with twin-engined Wellingtons, with smaller numbers of Whitleys and Hampdens. There was a growing force of the powerful Short Stirling (four Bristol Hercules sleeve-valve engines) and Handley Page Halifax (four RR Merlins), and these had even made a few tentative raids in daylight, mainly to bomb the German capital ships *Scharnhorst* and *Gneisenau* in French ports. Both these great bombers were rather difficult to make, whereas another large new bomber, the Avro Manchester, was a production engineer's dream. Unfortunately the Manchester's two big Rolls-Royce Vulture engines were unreliable, to the extent that one Manchester squadron, No 97, became known as 'the 97th Foot' because they were so often grounded. The answer was to fit longer outer wings with two of the reliable Merlins on each side. The resulting four-engined aircraft, the Lancaster, was a superlative night heavy bomber, first flown (as a rebuilt Manchester) on 9 January 1941.

Like all heavy bombers of the period it had a large crew, in this case seven, three of them manning power-driven turrets for aiming defensive machine guns. One of the unsolved mysteries is why these three turrets covered the nose, the upper hemisphere and the tail, whilst leaving the bomber's underside not only undefended but without even a single small window looking downwards. What makes this all the stranger is that by 1941 the Germans were experimenting with upward-firing guns on their night fighters, and by 1944 this *Schräge Musik* (jazz) type of installation was the preferred type on the newest versions of radar-equipped Ju 88 and Bf 110, used for shooting down 2,235 Bomber Command aircraft in that year alone. A further mystery is why, from late 1942, nearly all the Lancasters were fitted with a cumbersome mapping radar, code-named H_2S, which broadcast the presence and location of the bomber like an aerial lighthouse. Just to make the Luftwaffe's task even easier, most of the RAF 'heavies' were also fitted with a small radar whose purpose

Right: The 1944 Mustang was the P-51D, with a teardrop hood and was the last sub-type having a dorsal fin, (as Aircraft S from the 357th Fighter Squadron.)

was to warn of hostile night fighters. Its only practical result was to provide yet another source of signals on which the night fighters could home, from a distance of up to 72 miles (116 km)!

The dramatic Augsburg raid by Lancasters made the headlines, but hardly affected the war, sustained 60 per cent loss and provided the Germans with plenty of crashed Lancasters to examine. Bomber Command's basic style was the maximum-effort night attack, and in 1942 the targets had to be complete cities; even then, typically one bomber in five failed to find it. Really heavy raids opened with an attack on Cologne on the night of 30/31 May 1942 by 1,046 aircraft. Of these 599 were Wellingtons, and Harris scraped the barrel to assemble the force, using hundreds of bombers from OTUs (operational training units). Harris was misguided enough to think such a raid might bring 'abrupt termination of the war', and with the benefit of hindsight it can be seen that, though far heavier subsequent raids left Germany's cities in ruins, industrial production actually increased until the final collapse and occupation. Where Bomber Command did score was in its electronics, which added such valuable aids as Gee, a precision navigation aid which at last stopped the crews from being 'lost as soon as we leave the aerodrome', and Oboe, an even more precise aid based on ground stations which enabled a target to be marked by the élite Pathfinder Force using colour-coded pyrotechnics put down within a few metres of the aiming point, so that the follow-up Main Force could bomb with assurance.

From 1943 most of the Oboe-guided markers were Mosquitoes. This aircraft originated in a proposal by de Havilland Aircraft in October 1938 for a high-speed bomber built of wood, relying not on gun turrets for defence but on speed and height. The RAF and Air Ministry showed no interest, and even after the outbreak of war tried to force the company to drop such a heretical idea, and make wings for Halifaxes instead. By dogged persistence DH built a prototype, and when this flew on

Left: The bomber with the range to take the war to Japan was the mighty Boeing B-29 Superfortress. Here 'fifty-caliber' ammunition goes aboard.

Above: Barnes Wallis devised the 'Upkeep' spinning-drum depth bombs carried by the specially modified Lancasters of 617 Sqn to breach the German dams.

25 November 1940 it shattered all previous ideas about bomber performance. Much faster than any other British aircraft, despite seating the pilot and navigator/bomb aimer side by side, it could overtake a Hurricane with one engine shut down! Even more significant was the gradual realisation that the Mosquito was supremely versatile, and could do almost any task better than anything else available. Even the first batch included high-flying reconnaissance Mosquitoes and night fighters armed with machine guns and cannon, and equipped with radar. The first to enter RAF service photographed naval bases down the French coast as far as Bordeaux, and simply outpaced the Bf 109s sent to intercept it. The B.IV bomber version, carrying four 500-lb (227-kg) bombs, actually attacked Cologne on the morning after the first 'thousand-bomber raid' mentioned earlier. But what hit the headlines was the brilliant attack on the Gestapo (secret police) HQ in the busy centre of Oslo on 22 September 1942. Flying at below treetop height, and navigating by dead reckoning, the aircraft of 105 Sqn roared at almost 400 mph over the city centre and bombed with amazing accuracy. Unfortunately one bomb hit squarely and failed to explode, while three others went through the building and out through the far wall before detonating.

This was just the first of countless low-level raids by 'Mossies' (pronounced Mozzies) which told the Germans that nowhere in Europe was ever going to be safe again. Bombs fell on special parades in Berlin addressed by Nazi leaders, bombs were placed on individual parts of factories, and on 18 February 1944 eleven Mosquitoes breached the walls and outer wall of Amiens prison, despite a blinding snowstorm, freeing 258 of 700 French Resistance prisoners including many due to be executed on the following day. The most numerous mark of Mosquito was the FB.VI fighter-bomber, with four cannon, four machine guns and various loads of bombs or rockets. These scoured Europe looking for targets, and with Coastal Command hit hundreds of enemy ships. From 1943 the bomber and reconnaissance versions were chiefly high-altitude models with two-stage-supercharged Merlins and pressure cabins, with bulged bellies either to house extra fuel or a 4,000-lb (1,814-kg) bomb. It amazed the officials to discover the little Mosquito could carry a bomb of this size to Berlin, and towards the end of the war Berlin raids were suffering only one loss per 2,000 sorties – in sharp contrast to Bomber Command's heavies, where 10 per cent was common.

Not only did the bomber Mosquitoes suffer fewer losses but they were able to bomb with precision, whereas the Bomber Command heavies typically carried a giant 'blockbuster' and heavy boxes filled with incendiaries and let them go anywhere over a target city. There were two famous exceptions to this, one involving the Earthquake bombs and the other the Upkeep weapon, both of which were designed by B. N. (later Sir Barnes) Wallis, the former designer of the R.100 airship and the Wellington bomber. Earthquake was a giant bomb weighing 22,000 lb (9,980 kg) which needed a stripped-down Lancaster to carry it. Falling at supersonic speed, it penetrated far into the ground before exploding, the shockwaves then demolishing any structure nearby. On 14 March 1945 an Earthquake shook down six spans of the Bielefeld viaduct, which had resisted hundreds of lesser bombs. As for Upkeep, this was the unique spinning-drum mine designed to destroy the great German dams. Wing Commander Guy Gibson, one of the outstanding young leaders of Bomber Command, was picked by Harris to form a special squadron, No 617. They practised flying at very low level, and on 17 May 1943 they made their famous attack on the Möhne, Eder and Sorpe dams. Flying at exactly 60 ft (18.3 m) at 240 mph (386 km/h) they released the giant 9,250-lb (4,196-kg) bombs at just the right distance, the bombs having previously been spun up to 500 rpm in the opposite direction to their travel over the water. Skipping across the protective nets, they each sank down the face of the concrete dam until they were exploded by a hydrostatic fuse. Ever since, 617 have been famous as 'the Dambusters'.

Such precision attacks were unusual for Bomber Command, but they were central to the US concept of strategic airpower which was based on the ability of the heavily armed bomber, flying in large formations at high altitude, to penetrate enemy airspace and bomb accurately by day. By 1940 this had been practised many times by the first B-17 squadrons, using the Norden bombsight. The RAF's bad experience with these aircraft led to a major redesign of both the B-17 and the later B-24 Liberator, with much heavier defensive armament typically amounting to 13 heavy machine guns of so-called 'fifty' (0.5-in/12.7-mm) calibre, of which eight were in powered turrets. Such armament, with 500 to 600 rounds per gun, weighed many tons and needed a crew of five gunners quite apart from the two pilots, bombardier and radio operator. B-17s of the England-based US 8th Air Force made their first small combat mission on 17 August 1942. Gradually the 8th's strength mounted, and grew further with Bomb Groups equipped with the B-24, until massed formations even bigger than those of the RAF were flying out daily against targets in the heart of Germany.

Below: Production of the massive Republic P-47 just beat the P-51, and no fewer than 12,608 were built of the P-47D alone. Here a squadron is seen from an unusual angle.

Left: Mainstay of the heavy bomber wings of the mighty US 8th Air Force was the Boeing B-17 Fortress. The final model, unpainted, was the B-17G.

Losses, both in training and over Europe, were often grievous. The Luftwaffe, still flying the same old Bf 109s, Fw 190s, Bf 110s and even Ju 88 night fighters, fought doggedly and savagely, braving the hail of fire from hundreds of 'fifties' to try to bring down the American heavies. It was soon obvious that as well as taking out carefully selected targets with precision bombing, the massive raids by the 8th Air Force were also steadily taking out the fighter pilots of the Luftwaffe. Initial operations were escorted by RAF Spitfires, and then by USAAF P-38 Lightnings and also by the big Republic P-47 Thunderbolt, heaviest

single-engined fighter of the war, powered by a 2,000-hp Double Wasp with turbocharger and armed with no fewer than eight 'fifties'. These all had to turn back long before reaching the German frontier, but by 1942 various people – Ronnie Harker of Rolls-Royce in particular – had noticed what with hindsight seems obvious: that the P-51 Mustang would do better with a two-stage Merlin. The North American Mustang had been designed in 1940 solely for the British, who never thought of asking for the Merlin to be installed. First flown on 26 October 1940, the new fighter took full advantage of the latest technology in so-called laminar wing profiles (the thickness reaching a maximum much further aft than in earlier wings), low-drag cooling systems for liquid-cooled engines and overall good aerodynamic and structural design. At low level it was considerably faster than a Spitfire, despite having room for 180 gallons (818 litres) of fuel, compared with the British fighter's 85. It also had plenty of firepower, with four 'fifties' and four rifle-calibre guns. The only drawback was that the Allison engine fell off badly at high altitude, so early Mustangs and US Army P-51 and A-36 versions, were used mainly for low-level attack and reconnaissance. But in September 1942 the XP-51B with a Merlin showed sparkling performance at all heights. They joined the 8th Air Force in December 1943, and by 1944 were showing the amazing and unexpected capability of escorting the B-17s and B-24s all the way to Berlin and beyond, and then of jettisoning their drop tanks and slaughtering the Luftwaffe. Goering said "When I saw those Mustangs over Berlin I knew the war was lost".

The ability to cross Hitler's Third Reich by day with forces of over 1,000 heavy bombers was not gained easily, and at times it appeared a touch-and-go effort that took a heavy toll. But in fact the unequalled industrial might of the United States had long since been channelled into the next generation beyond the B-17 and B-24, the B-29 Superfortress. Launched in 1938, when Congress was denying funds for even the first production B-17s, the B-29 was in most respects the most advanced aeroplane of World War II. Indeed, many of its detail design features persisted in US- and Soviet-

Below: Though only a small number were built, the Welland-engined Gloster Meteor III was in action in 1944 with RAF No 616 Sqn, moving in 1945 to Belgian bases.

Left: Sir Frank Whittle's W.2B engine was produced in quantity as the Rolls-Royce **Welland. This highly successful engine was the first Allied turbojet to enter service (in June 1944).**

built aircraft into the 1960s. Planned as the ultimate strategic bomber then conceivable, it flew in prototype form on 21 September 1942, and after a gigantic effort production examples became available from July 1943 – though the first combat mission was not until a year later, over Burma. Among the many totally new features of the B-29 were large turbocharged engines of great complexity, pressurized crew compartments with the front and rear fuselage capsules linked by a tunnel, fully remote controlled gun turrets managed by various gunners at optical sighting stations who could if necessary override each other, a bomb load of up to 42,000 lb (19,051 kg) released from racks in front and rear bays alternately to preserve the centre of gravity, and a structure more highly stressed than anything previously attempted, with a wing loading of over 81 lb/sq ft (395 kg/m²) and wings skinned with metal up to 0.64 in (16.3 mm) thick!

For a year from summer 1943 this tremendously challenging aircraft was almost too much for its crews. Thanks to the vision and courage of US Army generals Westover and Arnold prior to 1940, it was the subject of a colossal production programme, by Boeing at Renton (Seattle) and Wichita, Martin at Omaha and Bell at Marietta (Georgia), each a vast windowless plant built in a few months specially to assemble B-29s. Many of the pilots came from B-17s who could hardly fail to notice that the 180 mph true airspeed at which the B-17 cruised was about the speed at which the B-29's special multi-ply tyres left the runway! Training missions in 1943–44 ran out of fuel after flying only half the expected distance, but this was a matter of training crews in the exact engine operating conditions for best fuel consumption. In June 1944 operational missions started, gradually increasing in scope and numbers until Japan was being laid waste in the most destructive aerial missions in history. On just one attack on Tokyo, on the night of 9 March 1945, vast areas of the city were burned and 84,000 died, another 100,000 being injured. The war was brought to an abrupt conclusion by two lone B-29 missions which dropped the first nuclear weapons, one on Hiroshima on 6 August 1945 and one on Nagasaki on 9 August.

To most Europeans the period after the German surrender on 8 May 1945 was anti-climax. In the final year of the war totally new weapons had rather suddenly come into action, each the result of ten or more years of development. These included the Messerschmitt Me 163 in May 1944, the Me 262 and Gloster Meteor in July and the Arado 234B in September, all manned jet aircraft, and the Fi 103 (so-called V-1) flying bomb in June and the A4 (so-called V-2) rocket in September. The last two were harbingers of a future kind of war to be fought at least partly by missiles, but the others were basically conventional aircraft propelled by a totally new form of engine which did not suffer from the fundamental limitations on flight speed which had previously been imposed by the piston engine and propeller. This new species of engine is called the gas turbine, made up of an air compressor, a combustion chamber where fuel is burned, and a turbine, driven by the hot gas from the combustion chamber. Such engines had a history going back to 1900 in practice, and to 1791 in theory, but all early examples had been thought of as a possible replacement for the piston engine in driving a propeller. Such an arrangement, today familiar as the turboprop, would offer advantages in smooth steady burning of the fuel, absence of reciprocating (to and fro) motion, and a generally higher rotational speed, so that for a given power the engine could be lighter and more compact. But it was a young cadet at the lately opened RAF College at Cranwell who, in 1928, had the vision to invent the turbojet.

Frank Whittle had the advantages of being a brilliant mathematician, good theoretical physicist and outstandingly gifted engineering designer – all unnecessary bonuses to a future pilot in the RAF. He wrote an essay on future aircraft propulsion which was so far over the heads of his tutors and superior officers that few realised it had any merit, but he did manage to get sent to Cambridge where he took an engineering degree and passed with high honours. He had absolutely no success trying to convince anyone in either the Air Ministry or the British aircraft industry that a turbojet was worth considering, and only after years of heartbreak did he at last manage to scrape up some capital and launch a private company called Power Jets in 1936. Thanks to the BTH engineering company he was able to build a prototype turbojet, a startlingly good engine running on kerosene, and started it on 12 April 1937. It was the first in the world to run, but continued lack of official interest meant that his work was overtaken by a number of engineers in Germany who, starting later, were by 1940 to have six major engine programmes and seven major aircraft, all funded by the government. The first turbojet aircraft to fly was the Heinkel He 178, a poor aircraft powered by a poor engine designed by Hans von Ohain; but it was still No 1, and it flew on 27 August

1939. Heinkel later made a twin-jet fighter but this was eventually cancelled.

Totally unlike everything else, the little Me 163B Komet rocket interceptor stemmed from tailless gliders designed by Dr Alex Lippisch. The DFS 194 rocket testbed reached 342 mph (550 km/h) in 1940, and the larger Me 163 streaked to 624 mph (1,004 km/h) a year later. In the air the production Me 163B Komet of early 1944 was sweet to fly, and its steep climb to very high altitude and speeds at great heights were of a totally new order. Pilots of the first combat unit, JG 400, could wait until oncoming US heavies were almost overhead at 30,000 ft (9 km); then, with a great whoosh, they took off and zoomed up to engage them with two 30-mm cannon. But the 163 had two faults which together made it more dangerous to its pilots than to its enemies. It had no landing gear except a single skid, which made it extremely tricky; it took off from a trolley which the pilot then jettisoned, often bouncing back and hitting the aircraft, while landing on the skid had to be judged exactly right or the aircraft would slew round or overturn. Worse, the two liquids used in the rocket engine exploded if they were allowed to mix, and many Me 163B pilots were never found after such an explosion had occurred.

In contrast the Me 262A-1a interceptor, with four 30-mm guns, and Me 262A-2a fighter-bomber with two 551-lb (250-kg) bombs, were basically ordinary twin-engined fighters of outstandingly good design, which, like the 163, were a dream to fly. Where they were different was that slung under the wings were two Jumo 004B jets each of 1,980-lb (900-kg) thrust, able to accelerate the big fighter to some 550 mph (885 km/h). Allied fighters found the 262 hard to handle, except when it was landing or taking off. But of 1,433 flown before the collapse only about 300 got into combat units; and they were not only unable to influence the war but the combat results showed that as many 262s were lost from all causes as the number of Allied aircraft they shot down. The Ar 234B was a rather larger twin-jet, with the pilot seated in the extreme nose, which was first used as an uninterceptable reconnaissance aircraft – which the Luftwaffe had desperately needed since 1942 – and by January 1945, as a bomber. Again, too few were available to have much impact. In the closing months of the war Heinkel's He 162, the Volksjäger (people's fighter), went into production at a rate planned to build up to 4,000 per month! This tricky machine, which experienced fighter pilots found difficult to fly, yet which was intended to be flown by 16-year-old Hitler Youths with only the briefest prior training, never quite got into action.

In complete contrast, the RAF's Gloster Meteor was a very ordinary twin-engined fighter, whose only odd feature was its revolutionary new Rolls-Royce Welland (later Derwent I) turbojet engines. In July 1944 No 616 had no difficulty converting from Spitfires, and they were soon shooting down flying bombs, but no special effort was put behind the programme. While the thrust of the Welland was 1,700 lb (771 kg), the development team under Dr S. G. (later Sir Stanley) Hooker very quickly produced the Rolls-Royce Nene, with a thrust of 5,000 lb (2,268 kg). Not only did this stagger the Germans just after the war, but it could be run for several hundred hours at a time without attention, whereas the German jets had a total life between major overhauls of between 18 and 40 hours only. But the steam had gone out of the British effort by 1945, and nobody in the aircraft industry even built an aeroplane to be powered by the Nene (the first to enter service was the very pedestrian Supermarine Attacker fighter, which reached the Royal Navy in August 1951). Other nations were more imbued by the idea that winning a war did not mean a slackening of effort, and these notably included the USA and Soviet Union. Both made the fullest and most immediate use of the Nene, as related in the next chapter.

Right: The dawn of electronic warfare, navigating a B-29 at high altitude on 19 May 1945 using the AN/APQ-7 blind-bombing radar.

A little of the joy and beauty of flight is captured in this superb picture of Lightning interceptors of RAF No 23 Sqn; one also senses the thrill that the layman would feel to be aboard. The pilots cannot take their mind off the job, but hold station.

7. COME FLY WITH ME

Right: The simple D.H.114 Heron was designed to carry up to 17 passengers on up-country routes where airfields were often mere dirt strips.

ALTHOUGH IT WAS NOT REALISED at the time, the 1950s were a real Golden Age in aerospace. Technical development leaped ahead as never before. At the start of the decade the speed of the fastest fighters was 600 mph (966 km/h); at its conclusion it had added a further 1,000 mph (1,610 km/h) and people talked more in terms of Mach numbers (the aircraft speed expressed as a decimal fraction of the local speed of sound). The speed of commercial transports rose from 300 to 600 mph, while the capacity of the biggest machines rose from around 50 to more than 200. Missile range increased from 200 to more than 6,000 miles (10,000 km), and space was conquered at last with artificial satellites which were clearly the harbingers of many other spacecraft yet to come.

It was the last decade in which it was taken for granted that aircraft performance would go on increasing, and the last in which the major industrial nations could accomplish whatever they wished in aviation. Subsequently the speeds of combat aircraft have, if anything, progressively fallen, and it has been realised that speed and height are no defence against air defences which can, however, be confused or misled by clever electronics and other countermeasures. Even more significant, the SST (supersonic transport), the first designs for which confidently appeared in 1956, proved later to be rendered commercially unattractive by the soaring price of fuel and the erection of political barriers. In a nutshell, inflation was to transform the scene utterly, so that each new aircraft development now has to last ten, 20 or 30 years, while many attractive new ideas today have to remain on paper for lack of money to try them out.

So different were the 1950s that Britain, perhaps not realising what time it was, launched 175 new aircraft projects between 1950 and 1955 (between 1975 and 1980 the corresponding figure was just five, all modest and two being versions of existing machines). The figure of 175, which had not dissimilar parallels in many other countries, stemmed from the rather sudden breadth of new technologies and possibilities, the dramatic opening up of horizons by gas turbine engines, guided missiles, large helicopters and many other new challenges, and also by the outbreak of a bitter war in Korea in June 1950 which removed for all time anything like the 'ten year rule' (which had been attempted by Britain's economy-minded government between 1945 and 1950) and ensured ample funding for any major new defence programme.

In fact continued high expenditure on defence had been assured from 25 July 1948, when the Soviet occupying authorities in Germany (today East Germany) abruptly closed all the road, rail and canal links to Berlin. Berlin, largely rubble, was still home to 2,500,000 people, and to the four-power Allied control commission. Stalin had no doubt that eventually his blockade would simply starve the Allies out, so that he could take over Berlin completely. He reckoned without the will and determination of the Western leaders. Instead of backing down, they organised the Berlin Airlift, made up of a mighty and growing fleet of C-47s (the mass-produced wartime DC-3s), big four-engined Douglas C-54 Skymasters (whose civil DC-4 version was also available in quantity), a motley collection of RAF Yorks, Hastings and

Left: Checker-painted vans house another invention to help in bad weather landings: GCA, ground-controlled approach. The aircraft are USAF C-54 Skymasters on the Berlin Airlift.

Right: An Airspeed Ambassador takes off from the factory at Christchurch. BEA's order for 20 was the only sale

Sunderlands, British civil Lancasters, Lancastrians, Haltons and Halifaxes, Tudors and Sandringham flying boats, and even the monster new Douglas C-74 Globemaster with a load of 27 tonnes or 125 passengers. At first the Russians fumed angrily, and threatened all manner of 'grave difficulties' if the airlift continued; they announced intense fighter exercises in the air corridors which linked Berlin with the West even though this was, for obvious safety reasons, expressly forbidden. Day by day they waited for the Allies' nerve to crack. Instead, a growing force of many hundreds of flight crews toiled ceaselessly to bring in food, gasoline, coal and people, and take other people out. Hundreds of shiny new airliners were stripped of their furnishings and droned day and night laden with anything from 150 sacks of coal to a flock of live sheep.

Not least of the achievements was to organize the airlift so that it could operate with safety around the clock. Nothing like it had ever happened before. The final air traffic control in the corridors (each 20 miles/32 km wide) was to space aircraft exactly three minutes apart, with 500 ft (152 m) vertical separation between successive machines, all traffic maintaining a cruising speed of 160 knots (184 mph/297 km/h). At first there were problems in bringing aircraft safely in to land in bad weather, but soon complete installations were made of the two pioneer bad-weather landing aids, ILS and GCA, as described later. The objective at the start was to airlift 750 tons (762,048 kg) a day, but this modest target was soon exceeded. By 1949, 8,000-ton (8,128,510 kg) days were common, and the record was 12,840 tons (13,046,260 kg) flown in on 16 April 1949. The sheer effort involved can be better understood when it is remembered that, by mid-1980s standards, the aircraft were puny; a single C-5 today could carry more than 45 average Berlin Airlift machines. And the effort paid off, the Russians at last permitting surface transport to continue on 11 May 1949. Western guts made *them* back down, instead.

This totally unnecessary episode, the result of a major miscalculation by the Kremlin of the nerve possessed by its former allies, played a major role in maintaining the Iron Curtain separating the East from the West, and in perpetuating the Cold War. Since then the population of many of the Eastern Bloc countries have been forbidden to visit any other countries. On more than 15 occasions the use of aircraft has enabled serving officers and the occasional civilian to escape, one of the more recent attempts being by balloon. One of the inevitable consequences of this sad division has been to sustain high defence expenditure in both East and West, but before going on to look at the consequences it is essential to return to the acronyms ILS and GCA, which were just two of the major advances made in electronic navaids (navigation aids) since the original form of Radio Range navigation on the civil airways in the 1920s. Both were important to Berlin.

During World War II a far greater number of aircraft than have existed at any other time were operated whenever the weather permitted. On the Eastern Front, in the Soviet Union, weather was often terribly severe, but navaids were sparse and, despite skill and courage by the air crews, the casualty rate was appalling by modern standards. In the West more than 6,000 aircraft crashed into hills or mountains (almost 3,000 in Britain alone), often within a few feet of the summit, causing a disgraceful drain on resources quite apart from the human suffering. By comparison, few landing accidents were caused solely by bad weather, though even here the figures were bad enough. The worst enemy was fog, and one of the few ways of trying to combat this was the British scheme called Fido, in which fuel was piped to hundreds of burners along the side of the runway to heat the atmosphere and evaporate the fog. Fido saved hundreds of four-engined bombers in the final two years of the war, at a handful of RAF Master Diversion Airfields, but it was crude and expensive.

The invention of radar enabled aircraft to be 'seen' on the ground on a CRT (cathode-ray tube) display very much like an early TV receiver. It was the germ of a possibility for an all-weather landing aid and by 1945 GCA (ground-controlled approach) was coming into general use at major airfields. It was based entirely upon one, or sometimes a pair, of ground radars sited close beside the runway to scan the sky in the direction of the landing approach. Each aircraft arriving at the airfield would be told to aim to join the glidepath – the landing approach path, sloping down at perhaps 3° to the threshold of the runway – at a considerable distance, typically 20 miles out. Inside the darkened GCA hut or caravan skilled operators would watch the incoming aircraft on two CRT displays, one showing it as a bright 'blip' (spot of bluish or greenish light) creeping down the glidepath seen in side view, and the other showing it head-on or from above. The pilot might be flying completely blind, unable to see the ground or have any idea of his position or altitude, other than the general assistance given by his instruments (which would be completely inadequate for a landing to be made), but the ground controllers could at all times see exactly how the landing was proceeding. All the pilot had to do was to set up the correct airspeed, rate of descent, engine throttle and propeller pitch positions, lower the landing gear and flaps, and then fly according to the GCA instructions. The ground controllers might say from time to time, "You are wandering to the left of glidepath . . . now you are slightly below the glidepath . . . you are six miles out, on glidepath . . . three miles . . . look ahead and land". When the final instruction was given the controllers would be certain that the runway, or its lighting, would have at last come into view, with the aircraft perfectly positioned for a successful landing.

Above: This Super VC10 is making an automatic landing guided by an ILS localizer similar to that in the foreground but at the other end of the runway.

Below: ASMI (airfield surface movement indicator) is high-definition radar used to picture traffic on airfields. This is a radar picture of London Heathrow.

GCA, using what later became known as PAR (precision approach radar), was the first method which enabled ordinary aeroplanes to land in very poor visibility. It was not adequate, except in emergency, to deal with dense fog; there remained a vital period of changeover from ground control to visual landing in the normal manner, though when there was no alternative many aircraft were 'talked down' right on to the runway. It had plenty of limitations, and always had the basic problem that it put the safety of the aircraft in the hands of controllers sitting on the ground.

ILS (instrument landing system) kept matters totally in the hands of the captain

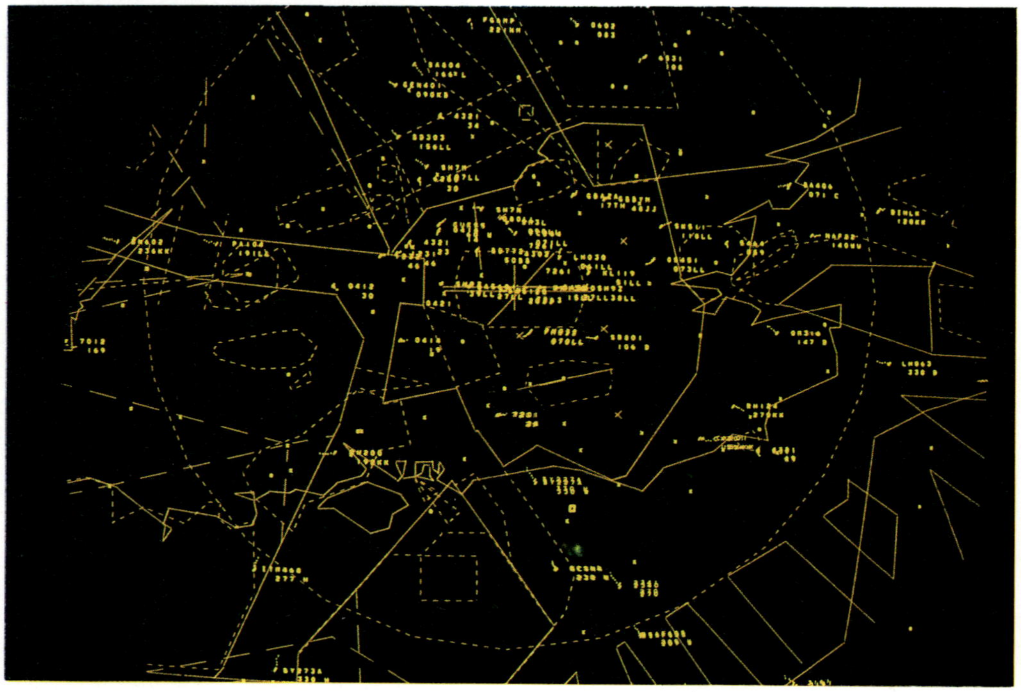

Left: Large civil surveillance radars give air-traffic controllers an overall picture of traffic, scheduled flights being automatically labelled. This is south-east England.

Right: For 35 years the worldwide electronic landing aid has been ILS (Instrument Landing System). This is the glidepath aerial (antenna) which provides guidance in the vertical plane.

Above: Inside the de Havilland factory at Chester in early 1958 showing production of the largely redesigned Comet 4. The four Avon engines will be fitted inside the roots of the quite thick wing.

of the aircraft, and instead of being a radar method used exactly the same principle as the old radio range in having pairs of overlapped radio beams, which in this case were sent out continuously, day and night, from transmitters near the so-called instrument runway. On the runway centreline, just beyond the far (upwind) end, would be a giant horizontal array of radio aerials (in North America called antennas) sending out two beams of radio signals each modulated at a slightly different frequency, and overlapping along the runway centreline. An ILS instrument in the aircraft cockpit tells the pilot exactly what proportion of each beam is being received; the pilot steers so that a particular needle in the instrument is centred vertically, and he then knows he is coming into land on the runway centreline. The ground station and its beams are called the localizer. Just beside the runway is another station which emits two vertically arranged beams called the glideslope. In the same way these give vertical guidance, and if the pilot keeps a second ILS needle exactly horizontal he knows he is on the glidepath, neither too high nor too low. ILS grew out of a US Army installation called SCS-51 (Signal Corps Set No 51), and like GCA it was in use by the end of World War II. It has gone on developing ever since, and today the very best ILS installations are so reliable and accurate that the captain can steer right down on to the runway in completely blind conditions.

Back in the pre-war era pilots could navigate using the radio range only by flying from one range station to the next, and this unfortunately implanted the idea of going from one fixed ground station to another in the minds of the Americans who, for obvious reasons of industrial and political power, have ever since in effect decided the world scheme for electronic air navigation. The first navaid other than simple radio to be carried in the aircraft was the D/F (direction finding) loop aerial. Mounted on a turntable above or below the fuselage, usually inside a streamlined fairing to protect the loop from damage, this indicated the direction to any selected ground radio station. The bearings of at least two stations were needed to get a 'fix' (accurate position of the aircraft). By the 1950s it was more common to rename the system ADF (auto direction finder) and make it drive a cockpit instrument. Instead of rotating a vertical coil the signals were received by two flat horizontal coils flush with the aircraft skin, which had to be cut away and covered with dielectric insulating material over the coils. In the case of the world's first jet airliner this seemingly small structural change was to have terrible consequences, as explained later in this chapter.

Back in the middle of the war the Gee navaid had at last enabled Allied navigators to find their way on the darkest night. Previously, in 1940, the Luftwaffe had sent out radio beams from Europe to guide its bombers raiding Britain, and had also been able to provide crossing beams or an additional signal giving 'pathfinder' aircraft a command to drop their bombs, but this was

Below: The world's first jet airliner, the prototype D.H.106 Comet.

useless for normal navigation. On the other hand, Gee could give accurate guidance anywhere in Europe, out to the range limits of the system. It relied upon four fixed radio stations in southern England, a 'master' and three 'slaves'. All sent out similar pulses of radio signals, in strict synchronization. If the pulses from one pair of stations were received by an RAF bomber over Europe at exactly the same instant, the bomber had to be on the line exactly bisecting at 90° the line joining those two stations. With all stations in use the navigator merely compared the times of reception of the various signals and read his position off a chart. In the USA a similar device was invented called Loran (long-range air navigation).

After the war the British Decca company refined the so-called hyperbolic systems Gee and Loran (called hyperbolic because the lines of equal time-difference on a map are hyperbolae) into the Decca Navigator. This used not pulses but CW (continuous-wave) radio signals and it gave better results right down to ground level. At first the aircrew had to take readings of a row of dial instruments called Deccometers, but soon an automatic plotting chart called a Flight Log was giving a continuous readout of aircraft position with a line drawn across a map. Decca and Loran became common aids throughout the 1950s on both civil and military aircraft, and in 1957 the ICAO (International Civil Aviation Organization) met to decide on a worldwide standard method. The only rival to Decca appeared to be a very retrograde method proposed by the USA called VOR/DME. The fact that this method was adopted because it was American appeared at the time to be a colossal blunder, especially to the British. One of the many reasons for this belief is that Decca could guide aircraft flying anywhere, whereas VOR/DME forced aircraft to keep to narrow overcrowded airways, just as the Radio Range had done. To travel along any direct route was extremely difficult, and there were many other drawbacks such as the fact that VOR stations could be used only within line-of-sight range.

VOR, standing for Very-high-frequency Omni-directional Range, was just a new form of the old Radio Range, its one new feature being that each station sent out signals that were coded depending on their direction of transmission – from north, through east, south and west and back to north – so that if the air crew tuned in to a VOR thought to be somewhere near they were immediately told its direction. For 30 years airliners throughout the USA, and for different shorter periods in all other countries, have had to fly from VOR to VOR, endlessly doglegging their way to the final destination and increasing the risk of collision by keeping all traffic within the same crowded routes. DME, distance-measuring equipment, adds a small radio in the aircraft which sends out very brief pulses about 20 times a second. These are picked up at a DME ground station, which is always in the centre of a VOR station, and this instantly sends back a slightly different signal which is picked up by the aircraft. The time taken for the round trip is measured accurately, and the DME instrument in the cockpit gives a numerical readout of distance to the ground station. Today VOR/DME covers the entire world. The best that can be said of it is that it works; but even in the 1950s anyone with foresight could see that far better systems could be devised and would have to come eventually.

Another big change in the 1950s concerned airfields. It will be recalled that the first airports with paved runways often had four runways looking like a Union Jack, giving eight possible directions of takeoff and landing. This was needed because early aircraft were sensitive to wind direction; an aircraft landing at 50 mph (80 km/h) can hardly ignore a wind moving at the same

Above: In a normal 'manual' approach by a commercial transport one pilot only will fly the aircraft. The other monitors the approach on the ILS indicator.

speed. During World War II the dramatic rise in aircraft weights and wing-loadings caused most of the grass fields to disappear and the runways to grow to a minimum length for any major airbase of 6,000 ft (1.8 km). There were well over 1,300 British bases, which mainly had tarmac runways, but permanent fields and major airports had runways of concrete which had to be made very much stronger and thicker in the 1950s with the coming of big jet airliners. Meanwhile the increased speeds at takeoff and landing, typically to values around 150 mph (241 km/h), diminished the importance of wind. Quite suddenly it was realised that for most purposes a single runway was sufficient, although it had to be much bigger than the wartime type. London's giant new airport at Heathrow, built on the site of the Fairey test-flying airfield, had been started with great triangulations of runways looking like a scaled up version of the wartime fields, giving six different directions. The decision was taken to abandon most of these runways, and today it is the world's busiest international airport with just two really good runways – one called 28L/10R and the other 28R/10L. It is easy to decipher this if it is remembered that to all pilots directions are measured in degrees from 000° (N) through 09 (short for 090°, or due E), 18 and 27 back to 36 (360°, which is the same as 000°). Thus 28L means the left-hand runway of two parallel runways aligned almost due west. One runway is 12,800 ft (3,902 m) long and the other 12,000 ft (3,658 m). Their width is 150 ft (46 m), so Orville Wright's first flight could have taken place across one of these runways.

Modern runways have a very comprehensive system of lighting, all of which is supplied by mains power backed up by emergency electrical supplies, and is not only totally weatherproof but must stand up to extra environmental conditions such as the impact of a giant high-pressure tyre or being struck by a stone blasted at it by a jet. Along the sides of the runway are white edge lights, visible from any direction; they give the pilot the essential perspective view he needs. The centreline is marked by 200-watt flush white lights visible from the landing or takeoff direction only. Stretching for a mile (1.6 km) beyond the downwind end of the runway are the approach lights, mounted on tall poles and cross-bars and sloping down roughly at the inclination of the glidepath so that the poles furthest from the runway may be 200 ft (61 m) high. Often they are not just bright white or yellow (sodium) lights but are sequence flashers which sweep continuously towards the runway. The crossbar lights give an indication of horizontal and may be 200 ft (61 m) wide. On each side short of the runway are red crossbars, but these change to white once the threshold is crossed, marking a touchdown zone 2,500 ft (762 m) long. The actual threshold is marked by a single continuous strip of white or green lights, which on the other side mark the bright red strip showing the end of the

Right: A British Aerospace 146, the world's slowest-landing jetliner, about to touch down on the same runway from which the first Comet prototype took off in 1949.

runway when the runway direction of use is changed, as sometimes happens if there is a strong wind which changes direction. Most important, the VASIs (visual approach slope indicators) are groups of special lights at each side of the touchdown area which show the pilot if he really is at the correct angle on the glidepath. The original form was called Drem lighting, from an RAF wartime field near Edinburgh, which in 1939 was fitted with red, amber and green lights inside carefully angled tubes pointing at the landing aircraft. Today's VASI has groups of red lights and groups of white lights. If the pilot sees only red, he is too low; all white and he is too high, but if he is on the glidepath he sees red above white.

World War II left almost the entire planet covered with good airfields. More than any other country, Britain had planned its commercial air routes around flying boats, and in 1952 even flew a gigantic new flying boat, the Saro Princess powered by ten turboprops. When this was planned during the war it had seemed logical. At the same time a monster landplane, the Bristol 167 Brabazon, was designed to fly the North Atlantic non-stop. This task seemed such a challenge that the Brabazon was given a wing of 219 ft 6 in (66.9 m) span, so deep that, as in the Princess, its mighty engines were buried inside it with only the propeller shafts projecting. Both aircraft took ten years to plan, design, build, fly and develop. By 1952 they were flying past impressively at the Farnborough airshow, but it was already obvious that technology had overtaken them – quite apart from the fact that their early Proteus engines were perhaps the worst turboprops ever made. It was a great relief when both projects were cancelled, but it reinforced the unhappy situation in Britain in which the government played the role of customer for commercial transports, leading to suggestions of scandal and waste when things ground to a halt.

On the other hand, even though the costs of developing a new airliner were seldom as great as £1 million, this was still too much for individual airlines or manufacturers to find. There had to be some government participation, and it says much for British foresight that, in 1943–44, the Brabazon Committee – named for its chairman, Lord Brabazon, who as J. T. C. Moore-Brabazon had begun flying in 1908 – had been appointed to get the nation back in the business of building civil airliners. During the war Britain had been only too glad to leave the provision of transports to the Americans, though in fact more than 2,000 transport aircraft were built in Britain, three-quarters of them intended as bombers but such poor bombers that there was only one possible job for them to do (examples included the Warwick, Buckingham, Stirling and Albemarle). What was needed were really good, modern, purpose-designed civil airliners, and great emphasis was placed on Britain's world lead in gas turbine engines.

The de Havilland company built the Dove light transport to carry on where the pre-war D.H.86 and D.H.89 Rapide family had left off. Where the ancestors had been fabric-covered biplanes, the Dove was a stressed-skin monoplane of singularly at-

Below: Contrasting Lockheed prototypes of World War II were the XP-49 fighter of November 1942 and (bottom) the first Constellation, of January 1943.

Right: TWA's Boeing 307, the first pressurized airliner, provides a backdrop for stewardesses in 1946 uniforms; the 307s too had been refurbished.

tractive form, and it became Britain's best-selling civil transport with 542 delivered by the early 1960s. The same firm bought the Airspeed company back in 1940, and this had the sad effect of denying any proper sales support to the outstanding 60-seat Ambassador, powered by two 2,900-hp Bristol Centaurus sleeve-valve engines. Britain's new airline BEA only bought 20. De Havilland said, "Let's forget such old machines, and put all our effort into the new jet-propelled Comet". Oddly, Britain suffered from a profusion of top decision-takers in aerospace who jumped to conclusions too quickly. Many thought that jets would instantly make piston airliners obsolete; nothing could have been further from the truth, although by 1950 the US dominance in this field was almost unassailable.

Britain was certainly right to concentrate on turboprops. It flew the Armstrong Whitworth Apollo and Vickers-Armstrongs Viscount and wisely picked the latter, which made its first experimental passenger runs as early as 1950. A beautiful machine with a pressurized fuselage and four pencil-slim Rolls-Royce Dart engines, it began life with 990-hp engines and 24 passengers, but grew to have 1,400-hp engines and 50 seats by the time regular BEA services began in April 1953. Passengers loved the effortless 300-mph (482-km/h) cruising speed, the absence of vibration, the high-altitude flight above most of the weather, and the giant elliptical windows which gave a marvellous view. Trans-Canada Airlines bought a big fleet, and forced the development of a version tailored to North American use. For the first time ever, a British aircraft company genuinely geared itself to study and meet the world market, and in 1955 Capital Airlines of Washington bought no fewer than 60. Vickers stretched the Viscount with 1,900-hp engines and up to 80 seats, and eventually closed the production line in 1963 at No 445. Vickers went on to build the much bigger and faster Vanguard, this time purely for BEA. Powered by four 5,000-hp Rolls-Royce Tyne turboprops, this had one of the 'double-bubble' fuselages with a cross-section like a figure-8, the upper lobe for passengers and the lower for cargo and baggage. Cruising at 425 mph (684 km/h) it was a fine aircraft, but apart from a small batch for Trans-Canada nobody wanted it; everyone had quite wrongly become mesmerised by jets.

Moreover, many said it was silly to design the Vanguard when Bristol Aircraft had spent the previous ten years producing a machine just like it, the Britainnia. Experience with the Brabazon helped Bristol create a superb airframe for this beautiful machine, which was one of the original wartime Brabazon Committee suggestions, a 32- to 36-seater with four Centaurus piston engines. It grew into a 90-seater with Proteus turboprops, and eventually seated 139 and could fly non-stop services from New York to Israel. It was just what the world wanted, and a whole generation ahead of all rival machines when it flew in August 1952. Unfortunately Bristol took too long to develop it, and at the eleventh hour a stupid and quite trivial problem with icing in the inlet ducts of the engines – which was never the slightest hazard, and could easily have been avoided – caused BOAC, the original customer, to refuse to accept the Britannia for two vital years, by which time the other customers had gone away. Even BOAC, which at last began carrying passengers in February 1957, realised too late it had crippled the prospects for a superb long-range aircraft which soon became known as 'the Whispering Giant' because of its near-silence. (Back in the war another Bristol aircraft, the Beaufighter, had been called 'Whispering Death' by the Japanese.) In 1955 the British Government and BOAC had the amazing shortsightedness to cancel the big Vickers VC7 long-range jetliner and say that the right policy was to concentrate on the Britannia. Then, having crippled the prospects for the Britannia, BOAC in 1956 bought a fleet of Boeing 707 jets, the American rival to the VC7!

Since before the war there had been a major faction in BOAC that preferred US aircraft, and in fact this preference at that time had a solid basis of better performance and reliability by the American machines, and far better operating economics. In the new post-war gas-turbine era it was a different ball-game, and BOAC perhaps failed to notice not only that the British airliners were technically ahead of all rivals but also that building them could become a giant high-technology industry if only the British companies could be given proper support and assistance by the national airlines in getting their products developed for the world market. Tragically, the one outstanding example of a British world leader, in which bold design was combined with fast timing and competent (or so it seemed) engineering, was designed to fall flat on its face just when it had the whole world at its feet, after it had demonstrated by years of impressive flying that the ultra-conservative world airlines could actually dare to buy this radical product. Indeed, by 1954 airlines all over the world were rather suddenly coming round to the view that they had better buy it in order to avoid losing traffic to their rivals. The aircraft was the de Havilland Comet.

This was yet another wartime Brabazon proposal, and the world's first jet airliner. In the initial stages in 1944–47 nobody quite knew whether or not to choose a radical tailless configuration, or swept-back wings, or put the engines at the back or in the thickened wing roots. Eventually it was decided to put four 5,000-lb DH Ghost engines in the wing roots, backed up by two Sprite booster rockets to assist takeoff in tropical climates (these were later left off). But should it aim for the Atlantic with six passengers and a few sacks of mail, or try to carry as many as 24 passengers? Fortunately the final design had a conventional fuselage seating 36, and pumped up to twice the differential pressure of any other aircraft in order to retain a comfortable interior at 40,000 ft (12.2 km), twice as high as any other airliner.

The DH.106 Comet prototype took off from Hatfield on 27 July 1949, and the first passenger service opened to Johannesburg on 2 May 1952. Airlines in general were intensely interested but, except for Air France, UAT and Canadian Pacific, were content to wait and see what happened. What they discovered was that once passengers had experienced the absolutely smooth and almost silent flight through the dark-blue sky of the stratosphere, to arrive in less than half the previous scheduled time, they simply did not want to fly in lurching, noisy piston airliners thundering and vibrating through the storm-clouds for twice as long. De Havilland built the Comet 2 with more efficient Avon engines, and then the even more impressive Comet 3 – faster, longer-ranged and seating 80 or more.

Even PanAm bought the Mk 3, but in January and April 1954 two Comets with full passenger loads vanished over the Mediterranean. It was eventually discovered their fuselages had exploded whilst climbing up to cruising height, weakened by the fatigue of repeated pressurization and depressurization. The aircraft whose wreckage was discovered had begun to crack at the corner of one of the ADF aerial cut-outs, but another tested on the ground burst open at the corner of a window. Amazingly, de Havilland had designed all these cut-outs as rectangles, which naturally suffer a stress concentration at the corners. The Comet was eventually re-designed and the Mk 4 went into service on the North Atlantic, for which it had not been intended, on 4 October 1958. Subsequently 74 Comet 4, 4B and 4C aircraft

Below: First flown in 1946, the DC-6 became the most successful long-range liner of its day, 175 being followed by 74 cargo DC-6As and 288 DC-6Bs.

Right: The prototype DC-7 seen here first flew on 18 May 1953, and combined the superb Douglas airframe with the complex Turbo-Compound engine of 3,250 hp.

were sold, all far more capable than the Mk 1; but the unnecessarily long hiatus crippled the programme which should have run to 1,000 aircraft.

Back in the early 1950s the powerful US industry had been uncertain how, if at all, to meet the competition of the Comet. The need to do so was lessened by the fact that virtually all the world's transports were now US-built, and though they were long-established basic designs they were proving amenable to continuing improvement. Boeing was the only loser; its pressurized Model 307 Stratoliner and giant 314A flying boat of 1939 had been short runs, and even the great 377 Stratocruiser derived from the B-29 had sold only 55. In contrast Lockheed's beautiful Constellation and Douglas's efficient DC-6 and -7 had sales in hundreds to airlines all over the world. The 'Connie' had originally been planned by Howard Hughes in 1939 for TWA, but when the first flew on 9 January 1943 it was painted olive-drab and had become a C-69 of the Army. Civil models waited until 1945, when their advanced features – pressurization, Fowler flaps, thermal deicing, turbocharged engines and reverse-pitch propellers to shorten the landing run – combined with an unprecedented 323-mph (520-km/h) cruising speed to merit the title 'Queen of the Skies'. In 1951 Eastern Air Lines was first to use the Super Constellation, stretched to seat 95 or more instead of around 65 passengers, and in 1953 KLM introduced the first version with Wright R-3350 Turbo-Compound engines in which the exhaust gas spun three turbines geared to the propeller shaft to increase power from 2,700 to 3,500 hp. Many USAF and US Navy Super Connies swelled the total by 1958 to 850, some of the later military models being packed with complex radars and electronics and doing an unsung job around the world until 1970. Best-seller of all was Douglas, which developed the wartime DC-4 into the Double Wasp-powered pressurized DC-6, stretched this into the 88-seat DC-6B, and then fitted the Turbo-Compound engine to produce the DC-7 of 1953. This in turn grew via the longer-ranged -7B into the DC-7C Seven Seas, which was the first aeroplane ever able to take on the North Atlantic as a routine trip in either direction.

Sales of hundreds of these familiar piston-engined machines blunted the apparent competition of turboprops and jets from Britain that, by the mid-point of the decade, had not taken a single sale from the US industry. Indeed, as late as 1956 BOAC was buying a fleet of piston-engined DC-7Cs to help fill the gap left by failure of the Comet. But in the longer term, turboprops and jets would clearly take over. Allison Division of General Motors, previously known only for its liquid-cooled wartime fighter engine, led the US race to perfect the turboprop, using slim axial compressors fed by a long curving inlet duct past a remote propeller gearbox carried 28 in (71.1 cm) in front of the rest of the engine on struts. No suitable jet appeared until in January 1950 Pratt & Whitney ran the first JT3. The great company had produced piston engines in World War II to power more than half the aircraft of the Allies, the monthly output exceeding 11,000,000 hp, but it was five years late getting into jets. Determined to 'leap-frog' past its competitors, the great Connecticut team built the JT3 turbojet as one of the first two-spool engines, with separate low-pressure and high-pressure compressors in series, each turned by its own turbine at the speed giving best efficiency. The advantage was that splitting the work of compressing the air between two independent compressor spools enabled the overall pressure ratio to reach record high levels, and this gave not only high thrust but also dramatically reduced fuel consumption. At last the way was open for high-speed jet aircraft with intercontinental range. The greatest pressure was for bombers, but the leader in this field, Boeing, was also determined to be the first with a commercial jetliner.

Above: Three of the first Boeing B-47Bs to join the USAF flying with the 306th Bomb Wing in the summer of 1951.

Below: Following these trials with the prototype YF-84F, the RF-84K reconnaissance aircraft entered service.

Rightly, the Americans had regarded the Comet as a bold pioneer effort that would never be more than an interim type, which would make little impact on the world market. When de Havilland stretched it and fitted later engines Boeing felt the competitive pressure rising, but never doubted its own ability to build a much more advanced aircraft. Its background rested on world leadership in big jet bombers, and it is at this point that attention must be turned to the post-war bomber scene, before returning to the Boeing 707.

Britain paid little attention to new bombers apart from paper studies, but the US Army ordered prototype jet bombers from Douglas (XB-43), North American (XB-45), Convair (XB-46) and Martin (XB-48), picking the B-45 Tornado as the production type which entered service in 1948, the year in which the US Air Force was formed as a separate blue-uniformed service. Boeing had also been given a contract for the XB-47, but the Seattle company courageously decided to lose a year applying a wealth of new aerodynamic data captured in Germany, which among other things showed that the sudden increase in drag as the speed of sound was approached could be postponed by sweeping back the wing and tail surfaces. Instead of jet

versions of the B-29, Boeing began to think of totally new configurations. At one time it considered four turbojets in the top of the fuselage, with the jets blasting past the fin, but eventually selected a graceful but structurally challenging arrangement in which the engines were mounted in external nacelles, or 'pods', slung well below and ahead of the wing on slim pylons.

The B-47, as it finally flew on 17 December 1947, had a giant fuselage of almost perfect streamline form, riding on a high wing swept at 35° carrying two twin-engine pods inboard and two single pods near the tips. Six engines were needed to lift the weight of fuel, all of which was carried in the fuselage along with the load of nuclear or conventional bombs. The two pilots sat in tandem under a giant fighter-type canopy, with the navigator/bombardier in the nose. In the tail were remotely

Below: The ultimate Super Constellation was the Lockheed L-1649A Starliner, here seen at Paris Orly in 1957. It was the longest-ranged piston-engined airliner ever built.

Below: Perhaps the greatest of all bombers, the Boeing B-52 soldiers on into the 1980s, more than 25 years beyond its planned retirement date.

Right: This Martin B-57B had a long career in the USAF. First produced in 1954, the B-57B was a revised tandem-seat US version of the British Canberra.

controlled guns. The landing gear comprised tandem two-wheel trucks on the centreline, with light outriggers under the inboard pods. Though a tremendous challenge to its crews, especially in 'coffin corner' where at maximum weight at high altitude the limiting Mach number and stalling speeds almost coincided, the B-47 Stratojet was such a winner that 2,242 were built, including 274 by Douglas at Tulsa and 386 by Lockheed at Marietta.

All were supplied to USAF Strategic Air Command, the giant nuclear deterrent force whose motto 'Peace is our Profession' explains their belief that deterrence can prevent war (and it has certainly managed to do so for almost 40 years). But even with inflight refuelling by the Boeing-developed method (using a telescopic but rigid boom controlled by an aerodynamic flight-control system by an operator riding in the tanker), the B-47 did not have the range to fly the global missions in which SAC was chiefly interested. The first bomber able to fly such missions was the gigantic Convair B-36, conceived in 1941 as a vehicle with which to bomb Germany after the collapse of Britain. The development task was so great, and the pressures on other Convair programmes so heavy, that the prototype did not fly until 8 August 1946. Fully armed production B-36B Peacemakers flew from July 1948, and the 383rd and last B-36 flew in August 1954. Biggest landplane of its day, the B-36 had a span of 230 ft (70 m) and was powered by six Wasp Major piston engines, each of from 3,000 to, ultimately, 3,800 hp, buried inside the vast wings, where they were fully accessible in flight, driving 19 ft (5.8 m) pusher propellers. Most also had four J47 booster turbojets in twin pods under the outer wings which could be closed over when not in use. Most had eight defensive turrets each with twin 20-mm cannon, and bombloads up to 84,000 lb (38,102 kg) could be carried. Some had enormous camera installations, and there was a plan to carry McDonnell XF-85 Goblin jet fighters inside the weapon bays for defence. Another scheme, FiCon (fighter conveyor) carried an RF-84F jet reconnaissance aircraft, which could hook on again after photographing a distant target. It was said the crews tore pages off a flight-deck calendar during missions, and certainly these colossal machines have never since been equalled for mission endurance.

Convair built a swept-wing jet development, the YB-60, but this lost to Boeing's completely new B-52 Stratofortress, first flown on 15 April 1952. This giant, the successor to the B-36, was planned with turboprop propulsion, because with early turbojets the fuel consumption was too high to fly the SAC missions. It was Pratt & Whitney's JT3 engine that changed the picture, and this engine was picked by the USAF, and later by the Navy (to replace the failed Westinghouse J40), with the designation J57. The B-52 was designed with eight J57s in four twin pods, and the engine began its career at a rating of 8,700 lb (3,950 kg) thrust per engine. By the time the B-52 entered SAC service in 1955 this had risen to over 11,000 lb (4,990 kg), and it then climbed to 13,750 lb (6,237 kg)

Right: A most unusual view of one of the Soviet Union's giant Tupolev turboprop aircraft, in the version known to NATO as 'Bear-D'.

Below: Known to NATO as 'Badger-F', this version of the Soviet Tu-16 has an Elint (electronic intelligence) pod under each wing.

with water injection. Each takeoff was excruciatingly noisy, with clouds of black smoke, and the gigantic wing flaps suffered fatigue cracking caused by the buffeting and impingement of intense noise. Successive B-52 versions were built at a high rate throughout the 1950s, and for what seems an unbelievable price of $6 million each. (In recent years more than $100 million has been spent on each B-52 just to prolong structure life and update the electronics.) The final models were the B-52G with an integral-tank 'wet wing' sealed to form a giant fuel tank, and with many other advances, and the B-52H, with TF33 (JT3D) turbofan engines, which extend the maximum range; one holds the world distance record, set by flying from Okinawa to Spain on 10–11 January 1962, at 12,532.3 miles (20,168.78 km). Altogether 744 of these great bombers were built; they were planned to be replaced in 1960, but no replacement will arrive until 1986 at the earliest, and many may still be in use in 1990. This is typical of the effects of inflation in the modern world, but it is doubly tough on the B-52, first because it was not designed for such a long life, and secondly because it was designed to carry a few nuclear weapons in the calm stratosphere and instead, since 1962, has actually had to carry heavy loads of conventional bombs and missiles and fly in the dense turbulent air at treetop height to try and stay under hostile radars.

Right: Painted silver, the first production Handley Page Victors for the RAF in 1956 were graceful aircraft, with exceptional speed and range capabilities.

SAC did deploy one other bomber in the 1950s, but this had only a brief career. In 1949 the USAF began to consider a supersonic bomber, and Convair flew the first B-58 Hustler in November 1956. There were many problems in the development of this aircraft, and also a certain amount of cheating; for example, the bombload and most of the fuel was carried in an external pod, as big as most bomber fuselages, which could be jettisoned over the target, thus making the basic aircraft smaller. The B-58 was built to the dramatic triangular tailless shape called a delta, with a structure mainly of stainless steel; not least of its difficult new features was that its four engine pods were fully variable, with sliding and pivoted inlets and nozzles to match the shape, airflow and shockwaves to all Mach numbers from zero up to Mach 2 (1,320 mph/ 2,124 km/h). In fact it could exceed Mach 2 for over an hour at a time, and at subsonic

speed could fly 5,000 miles (8,000 km) without inflight refuelling. The crew of three sat in individual capsules which could be ejected as sealed units in emergency. By this time other military jets had ejection seats, fired upwards (and very occasionally downwards) by explosive charges to get the occupant clear in emergency no matter what the speed or aircraft attitude, but a sealed capsule has been used in only one other aircraft, today's F-111 made by the same factory. SAC set many world records with the B-58, but limited range and high costs kept production to 116 and forced a withdrawal in 1970.

Britain flew its first jet bomber on 13 May 1949, and at first sight the English Electric Canberra looked too primitive and small to be interesting. It was when test pilot Beamont demonstrated that it could manoeuvre like a fighter, and certainly outclimb and outfly the second-rate RAF fighters of the day, that it was seen to be a very useful aircraft; it was especially welcome because of its simplicity and reliability, so it appealed to foreign customers. Not all the latter were third-world countries, because in 1951 the Canberra became the first British aircraft since 1918 to be adopted as a standard peacetime combat type in the USA. Martin made 403 of various B-57 versions under licence. Surprisingly, having taken so long to get started – the B-45 was in service before a British four-jet bomber got to the design stage – no fewer than four very similar types of British strategic jet bomber were flown, and three entered service. The fastest and most efficient programme was the Vickers-Armstrongs Valiant, powered by four Avons, of which 108 were delivered in 1955–58. Avro built the delta-winged Vulcan, the tailless deep-winged layout appearing to offer the best prospect of keeping within the ridiculous gross weight limit of 100,000 lb (45,360 kg) (in the event most Vulcans weighed more than twice this). Handley Page delivered the Victor, with one of the first T-type tails. The Vulcan and Victor, which duplicated each other, were noted for their advanced structures (with light yet rigid sandwich skins), neat bogie main gears which folded inside the thin outer wings, and their high all-round efficiency. Both were made in two generations, the Mk 2 having engines of over 20,000 lb (9,072 kg) thrust and

Left: The Victor's rival was the Avro Vulcan, a tailless delta. The first prototype (but not production versions) had a wing that was a pure triangle.

increased wing area. Victors were rebuilt as tankers, but Vulcans continued in service as low-level conventional bombers into the 1980s, the final remarkable missions being 8,000-mile (12,900-km) round trips to the Falklands in May 1982.

In the Soviet Union there was a lot of catching up to do, but by Herculean efforts, and the maximum input from the best foreign technology, a fair measure of parity was gained by the mid-1950s. Stalin had repeatedly asked for B-29s during the war, but when three landed in the Soviet Union a crash programme was undertaken to copy it. The Tupolev bureau dissected every part. Then a Tu-70 transport was built, using many actual B-29 parts, followed by a modified Tu-75 transport and finally, in July 1947, by the Tu-4 bomber. This was basically a B-29 with Soviet ASh-73TK turbocharged engines and a different armament system with ten 23-mm cannon. From this Tupolev developed the bigger Tu-80 and much larger and more powerful Tu-85. These provided the basis for the Tu-88 bomber with two very large AM-3 turbojets, and the Tu-95 with a giant swept wing and four NK-12 turboprops, each rated at over 12,000 hp. Both these monster engines were to be very important to the Soviet Union for the next 35 years, from 1950–85.

The Tu-88, which entered service as the Tu-16 (called Badger by NATO), was built

Above: The last prototypes of the new transport derived from the Boeing B-29. With more powerful R-4360 engines and a taller fin it became the C-97.

Below: The Myasishchyev M-4 has had a service career as long as its opposite number, the B-52. Here an M-4 'Bison-B' is escorted by a US Navy F-4.

in large numbers as a direct counterpart to the B-47. Developed for many missile-launch, reconnaissance, Elint (electronic intelligence) and other roles, it had a longer active life than the US type and was still widely used in the mid 1980s. Likewise the giant Tu-95, by far the world's fastest and biggest turboprop, entered service as the Tu-20 (Bear) and for many years bothered the Americans because of its global range. Again, it proved capable of much development and prolonged use, and many remain in use for numerous tasks including ASW (anti-submarine warfare). A third type, the Myasishchyev M-4 (Bison), has had an equally long career, but was built in smaller numbers; even with a huge aircraft powered by four AM-3 engines it was impossible to achieve the desired range, and so there was no direct Soviet counterpart to the B-52. Some M-4s have been rebuilt as tankers for air refuelling by the British probe/drogue method.

The two Tupolev bombers proved suitable for conversion as civil transports. The Tu-116 was almost a demilitarised Tu-95, and of the three built one was accepted into civil Aeroflot use as the Tu-114D and used for many long and fast flights. The Tu-114 had a much larger fuselage equipped for what in 1957 was the record number of up to 220 passengers. About 30 were built, and used for numerous record long-distance flights, records for speed and height with heavy loads, and also for regular Aeroflot service on all the longest routes, from 1961 onwards. As for the passenger version of the Tu-88, this was a great success and as the Tu-104 was built in large numbers and not only widely used but also led to all the subsequent Tupolev subsonic civil transports, the 124, 134 and 154. The 104 first flew on 17 June 1955, and because of its proven derivation was able to bring the head of the KGB to London Heathrow as early as 22 March 1956. From then until October 1958 it was the only jet airliner in service, though in fact it was not very fuel-efficient, even when the passenger seating had been increased from 50 to 70 and finally to 100. Range was typically 1,000 miles (1,609 km) with full

Below: Another type with a long career was the Lockheed Neptune (1945–84). Later models had underwing jet pods, as on this US Navy SP-2H, Vietnam, 1967.

80011
U.S. AIR FORCE

10293
U.S. AIR FORCE

BOEING 707-320B Intercontinental
N761PA
BOEING
707
TurboFan

Left: First flown in September 1943, the D.H. Vampire was a superb jet fighter. Production was assigned to English Electric at Preston in 1944 where this Mk 1 was built

payload.

The only other jetliner with a timescale similar to the Tu-104 was the French Caravelle, planned as a national programme in 1951 and flown by Sud-Est as the SE 210 on 27 May 1955. The French wisely minimised technical risk by using proven Avon engines, although these were hung on the sides of the rear fuselage in what was then a new arrangement. Fuselage diameter was the same as that of the Comet, so that the complete Comet nose and flight deck could be grafted on. In other respects the Caravelle was totally new and extremely well engineered, and though Air France spent three years on route-proving trials before taking passengers from 12 May 1959, the Caravelle sold well in progressively improved versions, the total being 280, a record for any European jet airliner apart from the Airbus.

One batch of 20 Caravelles was sold to the giant US airline United, highlighting the surprising continued lack of a rival US product. Timing a jet airliner proved difficult. In Toronto Avro Canada had flown a perfectly sound 50-passenger short-hauler, the C.102 Jetliner, in August 1949, only a month after the Comet, but nobody wanted it. This, combined with effortless worldwide sales by the piston-engined Convair 240, 340 and 440, took the pressure off any US short-haul jet; in fact, it was the competition of the Viscount turboprop that at last spurred Lockheed to offer a larger rival, the L-188 Electra (totally unrelated to the original Lockheed Electra) which gained large orders when it was announced in 1955. Powered by four 3,750-hp Allison 501 turboprops, this conventional machine seated up to 100 and was noted for its short takeoff and landing and delightful handling with anything up to three engines stopped. It entered service in 1959, but soon suffered puzzling structural failure in the air. The cause was more obscure than the metal fatigue of the Comet, but the Electra was very quickly modified and back in service. The fact only 170 were built stemmed from a mistaken belief that, even on short hauls, the jet had made the turboprop obsolete. One country where this belief was not shared was the Netherlands, whose famous Fokker company saw that the Dart turboprop would gain experience rapidly in the Viscount and would be ideal for a twin-engined high-wing machine to replace DC-3s. First flown on 24 November 1955, it gained the distinction of licence production in the USA (by Fairchild), and it was a US-built machine that flew the first service in September 1958. It weathered the 'propellers are obsolete' storm and remains in production in the mid-1980s, with sales approaching 800.

Lack of an engine precluded US competition in this market, but the emergence in the early 1950s of the JT3 did open the way to a US long-haul jetliner, much heavier, longer-ranged and faster than the Comet. Boeing had spent much time on various Model 367 studies, and at last in May 1952, a month after the first flight of the B-52, the company committed $15 million of its own funds to the Model 367-80. It had its eyes on two markets: the world airlines, for an 80–130 seater to fly sectors of about 2,500 miles (4,000 km), and the USAF, for a tanker/transport to replace the C-97 of which vast numbers were being made. The military counterpart to the Stratocruiser, the C-97 was capable but forced SAC's jet bombers to reduce height and speed for each inflight refuelling. Despite this, 888 were built, the last in July 1956. The USAF refused to commit itself on a jet replacement, so funding the 367-80 took a lot of courage. The prototype flew on 15 July 1954, by which time the company had renamed it the 707. Soon the USAF placed an order for 29 KC-135 tankers, and in 1955 permitted Boeing to offer a commercial model. Remarkably, because it is one of the costliest modifications a company can make, Boeing decided to increase the width as well as the length of the fuselage in the commercial 707, of which 20 were ordered by PanAm on 13 October 1955. To Boeing's chagrin, PanAm also ordered 25 Douglas DC-8s on the same day. Douglas had no military back-up and were taking an even bigger gamble.

Both aircraft had four JT3 engines in single pods, bogie main gears and generally conventional design. By 1957 the JT3 was giving 13,000 lb (5,900 kg) with water injection, enough for a great increase in weight so that PanAm opened 707 services with a non-stop New York to Paris flight on 26 October 1958. But both Boeing and Douglas had planned longer-ranged models powered by the JT4 engine, the DC-8-30 being unchanged in size but with gross weight raised from 211,000 lb (95,710 kg) to 287,500 lb (130,410 kg). Boeing not only raised the weight in its Intercontinental model to 316,000 lb (143,335 kg) but also increased the size of the aircraft.

There followed what became called 'the jet buying spree', almost every major world

Previous pages: Boeing delivered 732 KC-135A tankers to the USAF. Some are being re-engined with CFM56 turbofans.

Page 162 bottom: In Vietnam the Douglas A-1 Skyraider served with every US armed force, on almost every kind of mission including shooting down MiGs.

Page 163: Between 1954 and 1982 Boeing delivered 80 C-135s plus 962 larger 707s; this was the second 707-321B.

Right: The MiG-17 was produced as a major redesign of the MiG-15, though at a glance it looked similar. This is an afterburning MiG-17F of 1953.

airline buying either the 707 or DC-8. In 1958 competition from the Rolls-Royce Conway, fitted to a handful of both types, caused Pratt & Whitney to convert the JT3 turbojet into the JT3D turbofan, a dramatic and relatively simple modification which many airlines did for themselves. This increased thrust, reduced fuel consumption and noise and even further reduced inflight failures, as well as eliminating water injection and smoke on takeoff. With the JT3D Douglas (McDonnell Douglas from 1966) developed the DC-8 Super Sixty series with even more fuel and bodies stretched to seat up to 251 passengers, finally delivering the 556th and last DC-8 in May 1972. Boeing did even better, having produced 732 KC-135 tankers and many other military versions – some of them grotesquely modified for secret electronic missions – as well as the short-haul 720 and a variety of 707s, the last of which was not delivered until March 1982, bringing the total of all 707-related machines to almost 1,800.

The so-called 'big jets' (though later dwarfed by today's 'wide-bodies') transformed civil aviation. They enabled many more people to be carried at twice the speed, over much greater distances and at fare prices lower than before, whilst at the same time eliminating the need for government subsidies around the world. This was just the opposite of what many had predicted, the stampede to buy jets having been widely criticised as a lunatic rush into bankruptcy. Another effect of the 707 and DC-8 was to force every country to upgrade its international airport(s) with much longer and stronger runways.

In many countries the single international airport is also the main air force base. Most such nations are in the Third World, and these were naturally the last to switch to military or civil jets, and mostly completed the 1950s with ex-World War II aircraft. In contrast the leading industrial countries often thought the piston engine was obsolete when it was still the best choice. An outstanding example was the US Navy Douglas AD-1 Skyraider, first flown in 1944 as a carrier-based attack bomber. Great efforts were made to build the turboprop XA2D Skyshark, which failed; meanwhile, not only did the old Skyraider prove capable of flying almost every kind of combat mission, so that it stayed in production until 1957, with

Below: First flown as the D.H.110 in 1951, the Sea Vixen reached the Royal Navy as a carrier-based interceptor in 1959. This is a later Sea Vixen FAW.2.

Left: Two RAF ground staff stand on the ladders to assist the instructor and pupil seated side-by-side in a Lightning T.5 interceptor trainer at RAF Binbrook.

Below: USAF counterpart of the Lightning, the Convair F-106A Delta Dart had an equally long front-line career and a few still fly with the US Air National Guard.

Right: This Vampire jet-fighter cockpit of 1950 would be familiar to any World War II pilot but looks archaic today. In the centre are the six basic blind-flying instruments.

3,180 built, but in the late 1960s in Vietnam it was realised that even this total was still not enough! One of its missions was AEW (airborne early warning), carrying a giant radar aloft to keep watch on everything within a radius of 200 miles (322 km). Another pair of piston-powered machines that stayed in production through the 1950s were the Grumman S2F Tracker carrier-based ASW (anti-submarine warfare) aircraft, and its big shore-based counterpart, the Lockheed P2V Neptune; later Neptunes did, however, have two booster jets to allow operation at increased gross weight, as did the RAF's even bigger Avro Shackletons developed for the same maritime patrol missions from the wartime Lancaster, via the late-1940s Lincoln.

Fighters, on the other hand, became almost 100 per cent jet as soon as the war was over in 1945. The leading engine was the British Rolls-Royce Nene, but amazingly this was almost ignored in its own country, and in September 1946 a batch was shipped to the Soviet Union. Here German swept-wing aerodynamics were already fully incorporated in new fighter designs, and the Mikoyan/Guryevich (MiG) bureau very quickly designed the I-310 to use the imported British engine. It proved to be an absolute winner, going into large-scale production from 21 March 1948 as the MiG-15. In the Korean war, started by the North Korean invasion of South Korea on 25 June 1950, the MiG-15 achieved an easy mastery over all Allied types except for the North American F-86 Sabre, one of the two US types deliberately delayed to incorporate swept-wing data (the other was the B-47). Powered by one of the totally US-designed axial jets, the General Electric J47, the F-86 first flew on 1 October 1947 and soon showed that in a dive it could exceed the speed of sound, making the then-mysterious sonic bangs all over southern California. Various F-86 models remained in production throughout the 1950s, the total easily passing the 9,000 mark. The chief single type was the F-86D all-weather interceptor, one of the first single-seat fighters with large radar, in this case with guns replaced by a battery of rockets fired automatically under computer control. Similar automatic 'collision-course' fire control was a feature of the Lockheed F-94 Starfire, Northrop F-89 Scorpion and Avro Canada CF-100, all of which were large two-seaters. Britain did not adopt this; the primitive Gloster Meteor and DH Vampire and Venom night fighters, each with four 20-mm cannon, being succeeded from 1956 by the Gloster Javelin, a very large delta twin-jet with four 30-mm guns and later with air-to-air missiles (AAMs).

Germany had almost put AAMs into service in 1944, but the first to go into regular service were the US Navy's Sidewinder, with infra-red homing on to hot parts of target aircraft, the Sparrow, with radar guidance, and the USAF's Falcon which could use either method. Undoubtedly the most advanced interceptor of the era was the Convair F-102 Delta Dagger, with an even later semi-automatic Hughes radar fire control tailored to Falcon missiles. First flown on 24 October 1953, it proved unable to meet its contractual commitment to fly on the level faster than sound. A period of frantic redesign followed, and the reshaped F-102A at last went into production in 1955. The first fighters able to fly at over Mach 1 in level flight were the North American F-100 Super Sabre and the Soviet MiG-19, both of which – after serious technical delays – were in production in 1955. Britain abdicated from the supersonic race in 1946, but the USA continued with its Bell XS-1 (later X-1) series of research aircraft, Capt (later Gen) Charles E. Yeager exceeding Mach 1 on the level on 14 October 1947. Subsequently, later X-1s and the US Navy Douglas D-558-II Skystreak both exceeded Mach 2, while on 27 September 1956 a Bell X-2 achieved Mach 3.2.

Unlike any period before or since, the decade of the 1950s was a time of colossal thrusting ahead, so that it seemed quite natural to design such fighters as the F-108 Rapier to fly at Mach 3 and the XF-103 to fly at Mach 3.7 or 2,446 mph (3,936 km/h). Neither reached the flight-test stage. Avro Canada's CF-105 Arrow was the biggest, fastest and longest-ranged of all fighters of the 1950s, but tragically this was cancelled by the Canadian government in 1959 in a short-term bid to save money, and replaced by a complex American SAM (surface-to-air missile) system that proved almost useless and was soon deactivated. In taking this decision the Canadians were strongly influenced by Britain, which in 1957 had announced that it had been decided that military aircraft were henceforth obsolete and would be replaced by missiles! All the many new fighters and bombers were cancelled, save only the Lightning for the RAF (which, it was explained "has unfortu-

led by the giants of Piper, Cessna and Beech thought, as did their predecessors in 1919, that with a world full of ex-wartime pilots almost everyone would soon have his own aeroplane – or even helicopter, because at last the helicopter had been turned into a practical proposition during the war. Piper built 36,000 lightplanes in 1946, but nothing remotely like this happened again, and the idea of a plane in every garage remained a pipe-drean.

On the other hand, sales of private owner aircraft stayed at far above the pre-war level, and they have grown ever since. Formula air racing grew from its beginnings in 1948 to encompass hundreds of aircraft all built to fly as fast as possible with a 100-hp piston engine. Aerobatics became an organised international sport in the 1950s, and so did the keen competition to put on the best show by military (air force and navy) formation teams, often equipped with the latest fighters but in striking paint schemes. In a nutshell, the 1950s were a time of unprecedented progress.

Above: Inside the former Blackburn factory near Hull, as Hawker Siddeley Buccaneer attack aircraft are produced for the Royal Navy. Later it was employed by the RAF.

nately gone too far to cancel") and the Buccaneer attack aircraft for the Royal Navy, which after much argument was allowed to continue because long-range missiles then found it hard to fly reconnaissance missions and attack such mobile targets as hostile fleets. Meanwhile in France the courageous Marcel Bloch, one of the top planemakers of the 1930s, changed his name to his Resistance *nom de plume* of Dassault and built the delta-winged Mirage which exceeded Mach 2 in 1958. Freed from competition from Britain, it was to catapult France into the No 1 spot among planemaking nations in Europe.

As for small private-owner types, these were by 1950 just one facet of a vast group known collectively as General Aviation. Other branches of GA are agricultural aircraft, business and executive types, aerial work (photography, survey, police, emergency services and private charter) and club machines and trainers. Gliding and parachuting grew fast in the 1950s, but the vast cult of the 'home-built' and the various microlights and hang gliders did not yet exist. The established lightplane builders,

Right: F-100A Super Sabres on the flight-line at NAA's plant at Inglewood in 1954. These aircraft later had to be urgently fitted with a taller fin giving better stability.

Powered by six General Electric J93 turbojets of 31,000 lb (14,062 kg) thrust each the two North American XB-70 Valkyries were the most powerful aircraft ever built. At full speed of 2,000 mph (3,219 km/h) (Mach 3) the outer portions of the giant stainless-steel delta wing were folded downwards.

8. THE EXECUTIVE JET SET

Left: The flight deck of the Hawker Siddeley Trident 3B, almost the last to be built of this type, is nevertheless somewhat dated today. The latest jetliners have large multicolour TV-type displays instead of traditional dial instruments, and plenty of keyboards for talking with on-board computers.

BY 1960 AVIATION was on the up and up, in every category and in every country, except for the United Kingdom which had decided that manned combat aircraft were not even to be mentioned. This caused a colossal hiccup in the British industry, and the hiccup was compounded by government-enforced mergers between the manufacturers which saw almost all the famous old names vanish, to be replaced by just two giant groups: Hawker Siddeley Aviation, which embraced Hawker, Gloster, Avro, Armstrong Whitworth, de Havilland, Blackburn and Folland; and British Aircraft Corporation, which was formed by Vickers-Armstrongs, English Electric and Bristol, and also took over Hunting. Westland, the helicopter builder, took over Saunders-Roe, Fairey and the Bristol helicopter division. In the same way, the nine British engine companies were forced into two groups, Rolls-Royce and Bristol Siddeley, and Rolls-Royce then bought Bristol Siddeley in 1966.

This was not all, traumatic though it was. Thanks to the incredible decision to stop making military aircraft from 1957 the British industry was placed in an almost impossible position, and increasingly the country came to rely on foreign products. After great arguments, and partly because of the far-sighted vision of British engineers who saw the future must lie with jet V/STOL (vertical or short takeoff and landing) aircraft, new military machines were at last permitted to be developed. For reasons apparently not connected with the aircraft, the Labour government elected in 1964 decided that building aircraft in Britain was generally undesirable; it called the plane-makers "overgrown and mentally retarded" and claimed that vast sums could be saved by scrapping British military aircraft (again) and relying on American types. Naturally this made the 1960s even more traumatic and controversial than the 1950s had been, and further eroded the nation's competitive position. In 1967 the government even pulled out of the European Airbus project, but by good fortune Hawker Siddeley had the sheer courage to stay in the programme as the partner responsible for the wing, using its own money throughout. This enabled a superb aircraft to be built on time, and kept Europe (including Britain) in the business of building commercial transports.

Curiously, the one big jetliner programme launched in Britain with government funds in the 1960s was the Concorde SST (supersonic transport) which proved to be a major commercial mistake, unlike the Airbus which from the start was an obvious winner. Back in 1960 it was not easy to see which direction to take. To a company like Boeing there is less of a problem; intimate relationships with almost all the world's airlines leads as a semi-automatic process to the emergence of a demand for a new type of jetliner, and it is then not too difficult to get the size, market and timing as refined as possible before going ahead. In the greatest possible contrast, the equally competent British manufacturers were almost forced to rely on the combined sponsorship of the British government and nationalised air corporations, and for the past 40 years this relationship has proved disastrous. A high peak of ineptitude was reached in 1955 when BOAC, the larger of what were then two nationalised British airlines, said it had no interest in big jet airliners; it said "The Comet 4 and Britannia will meet all our needs well into the 1960s". So Vickers had to abandon the VC7, the civil version of the V.1000 for the RAF which had been cancelled within weeks of its first flight. The need for a big jet was obvious to any bright schoolboy, and within nine months BOAC ordered a fleet of Boeing 707s!

Having thus handed the world market to the USA on a plate, BOAC then decided it might be a nice idea to build a big jet in Britain (presumably to replace the cancelled VC7). An order was placed for the de Havilland 118, but this remained a paper project. Instead BOAC asked Vickers to build the VC10, generally similar to the 707 but with rear-mounted Rolls-Royce Conway engines. A special feature was to be the ability to operate from the short runways on the BOAC routes to Africa, the Far East and Australia. Naturally, this means a bigger wing with advanced high-lift systems, which penalises the cruising performance and economics in comparison with an aircraft, such as the 707, needing a long takeoff run. Despite the need to pay for all the design and development the order for 35 was priced at £68 million, or £1.94 million each, whereas two years earlier the price for the 15 BOAC 707s was £44 million, or £2.93 million each, even though this aircraft was already designed. Eventually BOAC asked for a much longer Super VC10 as well, then it cut the size of the Super, and then it completely changed the numbers on order four times. Vickers managed to fly the first VC10 on 29 June 1962, and from the start both this and the heavier, longer-ranged Super VC10 proved outstandingly good and popular aircraft. There were just three small problems, typical of the British scene. First, the short runways had by this time all been extended to cope with 707s, as anyone could have foreseen. Secondly, by insisting on short field performance BOAC guaranteed that the VC10 could not equal the economics of the 707; it then publicly criticised the aircraft and did all it could to damage the programme and claimed government compensation for operating it. Thirdly, and not surprisingly, few other VC10s were sold and Boeing and Douglas never noticed any competition.

In parallel the short-haul airline, BEA, realised just too late that its big turboprop Vickers Vanguard ought to have been a jet. First flown in January 1959, the Vanguard was almost a reinvention of the Britannia but with more passengers and less fuel. Almost as soon as it placed the order in July 1956, BEA saw that passengers would prefer to fly by jet, and Lord Douglas, the airline's chairman, was forced to admit that "a few short-haul jet aircraft" might be needed to back up the main fleet of turboprops. By this time the three US giants, Boeing, Douglas and Lockheed, were all overworked. It was the ideal time for the British industry, shattered by the news that military aircraft were extinct in Britain, to go out into the world market and use its unrivalled experience and unrivalled engines to build the world standard new short-haul jet. Sadly, the relationship between the industry, BEA and the government made this a pipe-dream. Instead BEA wrote an exact specification, while the government announced that the order would only go to a merger of several companies. The candidates eventually narrowed to two: the Bristol 200, offered by Bristol, Hawker Siddeley and Shorts, and the de Havilland 121 offered by DH, Fairey, Hunting and Saro (the last-named dropped out). Although on paper the Bristol looked better and even interested PanAm, BEA picked the DH.121. By April 1958 this had settled down as a modern trijet with three 14,000-lb (6,350-kg) Rolls-Royce engines at the tail and carrying 111 passengers for up to 2,073 miles (3,336 km). This would have suited the world market and sold in very large numbers, especially if de Havilland had, against the odds, managed to persuade BEA to make it fit smaller runways than those on the corporation's trunk routes.

Suddenly BEA found that traffic was not rising as fast as they had forecast. It was still rising, and the 1959 slackening was only temporary, but the airline panicked and instructed de Havilland to cut the 121 in size to 97 seats over the much shorter range of 930 miles (1,500 km). This forced Rolls-Royce to design a new engine, the Spey, of 9,850 lb (4,470 kg) thrust. The redesign lost six months, but more significantly it resulted in an aircraft quite unattractive to the world market. At this time, in 1959, Boeing at last began to firm up its own ideas for a superior short-haul jet and the 727 came out almost identical to the original form of DH.121. It looked to the Seattle company as if it would have to share the market, but the amazing decision to cut down the 121's size and range was then followed by the order by Sir Aubrey Burke, de Havilland managing director of the so-called Airco consortium producing the 121, that his salesmen should on no account even speak to foreign airlines until the 121 had been developed to BEA's requirements! A few weeks later the Airco consortium was dissolved, de Havilland joining Hawker Siddeley, Hunting BAC and Fairey Westland! Lord Douglas of BEA then suggested de Havilland and Boeing should 'get together'; Boeing sent a top team to Hatfield (from where the British press were excluded) but the return trip was less fruitful, Boeing saying there was no reason for both companies to be stupid. In any case, Boeing did not need to copy Hatfield; with launch orders for the 727 of 40 for Eastern and 40 for United it could now see the start of a big programme.

The first DH.121, or Hawker Siddeley Trident, flew on 9 January 1962. By this time it was obvious even to BEA that it needed more capacity and range, and the Hatfield team spent the entire 1960s trying to restore what had been cut out in 1959, finally even having to add a fourth engine to get it off the ground with the necessary extra fuel and passengers. Altogether 117 Tridents were sold; excluding BEA and the Peoples' Republic of China, the total sold was just 17. This was the inevitable outcome of the crippling form of government/corporation procurement process preferred in Britain. In contrast the 727 remained at all times responsive to the world market, growing in capacity and range and ending

Right: This Rolls-Royce Spey, quite a small turbofan engine of 9,850 lb (4,468 kg) thrust, is about to be installed in the tail of the second Trident 1 for BEA.

Below: This DC-9-30 series is one of 19 operated by the Spanish carrier Aviaco. Though longer than the first DC-9s, the 30-series are dwarfed by the MD-80s.

Below: The 737 has matured as a smash hit, with over 2,000 sold. In 1984 the 737-300 introduced a new-generation engine, the CFM56.

Below: Though production in Britain terminated at a figure of 230, the BAe One-Eleven is still being built in Romania. This is one of the Romanian One-Elevens.

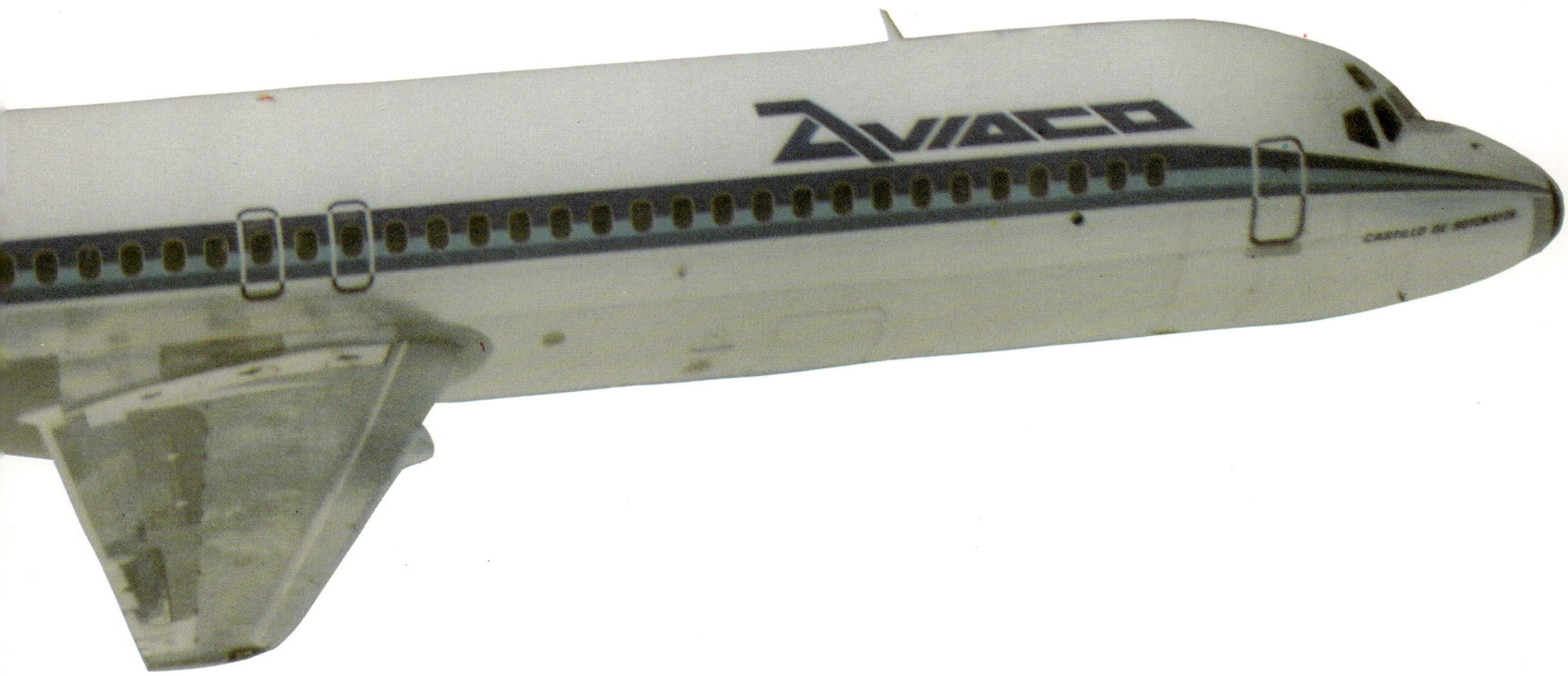

Below: The best-selling Boeing 727 had no government finance yet kept a Boeing plant at Renton busy making 1,832.

Above: This HS.125-3B was at one time used by the Australian overseas line for pilot training.

up at a gross weight of 209,500 lb (95,027 kg), compared with 115,000 lb (52,165 kg) for the first Trident. Boeing sold 727s to considerably more than 100 original customers (others have bought second-hand), and the total reached 1,832, more than for any other commercial transport in history. (The DC-3 total was many times greater, but most were built under military contracts.) As a footnote, BEA followed BOAC's lead in publicly criticising the aircraft that had been tailored so exactly to its own stupid specification, claimed government subsidies for having to operate it, and in 1967 tried to buy 727s!

What is especially galling is that British design engineers are unsurpassed, and when the management is able to take reasonable decisions the result is likely to be competitive. When BAC was formed in February 1960 its smallest partner, the former Hunting Aircraft – one of the unhappy Airco partners on the DH.121 – had been trying to sell a small twin-jet called the H.107. This was an attractive design with the newly fashionable rear-engined T-tail layout. BAC enlarged it to fit two of the new Rolls-Royce Spey engines and the resulting machine was offered as the BAC.111, or One-Eleven. British companies seemed to have difficulty thinking of memorable numbers, and as 111 was often written III it became common in the USA for this excellent machine to be called the 'Back three', and introduced as such by the cabin staff to the enquiring passengers. That there were such passengers stemmed from the fact that the basic design was right, even though BAC needed plenty of nerve to go ahead on the basis of a single order for 10 from the independent British United, whose boss was Freddie Laker. Because of the limitations of the Spey BAC could not stretch the One-Eleven much, but it grew from 79 to 119 seats and in weight from 73,500 lb (33,340 kg) to 104,500 lb (47,400 kg). First flown on 20 August 1963, from Hurn, near Bournemouth, the One-Eleven eventually reached a sales total of 230, about half in the USA. In 1979 British Aerospace signed a deal with Romania for the transfer of One-Eleven production to that country, and this big collaborative deal is now in full operation.

Substantial sales of One-Elevens to American, Braniff and Mohawk in the USA in 1963 made it inevitable that a US competitor should soon emerge. This materialised as the DC-9, very similar to the One-Eleven but basically larger because of the greater power of the Pratt & Whitney JT8D engine (which was almost identical to the engine planned for the original DH.121 before BEA made it smaller). Thus the DC-9 began at 77,000 lb (34,930 kg) with some 72 seats, and its certification programme confirmed the ability of the strong US companies to move faster than the British, the first flight being on 25 February 1965, followed by two more DC-9s in May, one in June and another in July, to complete certification in November. For comparison, the One-Eleven, fastest of all British jet certifications, took from August 1963 until April 1965, so the DC-9 almost caught up with the British jet at the very start. Douglas saw that the market needed different sizes, and over a 15-year period rang the changes on the DC-9 to a greater extent than has been done elsewhere. The smallest model, the DC-9-10, seats 65–72, while the Dash-30 took capacity to 115, the -40 to 125, the -50 to 139 and the largely redesigned -80 to 172, all with identical body sections and the same basic wing. It was obvious that the DC-9 would sell quickly, and this proved such an embarrassment to Douglas that the company became financially over-extended. This was despite sewing up several big cost-sharing agreements with suppliers, the latter bearing the risk by paying for the costs of tooling and materials and Douglas undertaking to buy a guaranteed minimum number of parts for not less than a stipulated price. For example, de Havilland of Canada, originally a British company, contracted to make the wing, rear fuselage and tail, and eventually became a subsidiary called Dacan, Douglas Aircraft of Canada. Such agreements have become increasingly necessary in order to get astronomically costly new aircraft projects off the ground, yet Douglas still ran out of money and eventually, on its bankers' recommendation, had to merge with a military planemaker, McDonnell Aircraft of St Louis. McDonnell Douglas Corporation was incorporated on 28 April 1967. It started with

Above: 'The missile with a man in it', the Lockheed F-104 Starfighter was built for flight performance at the expense of anything else.

Right: First of the production executive jets, the Lockheed JetStar was also one of the biggest. In 1975–80 Lockheed built a further 40 with TFE731 fan engines.

140,000 employees, sank to 57,800 in the tough mid-1970s and today is stable at 72,000.

Today the total sales for the DC-9 and its successor, the MD-80, easily exceed 1,200; without competition the total might be double this. Boeing was convinced the market for short-haul jets was big enough to support a rival, but in the mid-1960s all the US carriers had either bought DC-9s or were just about to. The exceptions were two giants, Eastern and United. In 1964 Boeing firmed up a proposal for the 737, an attractive but rather portly twin-jet with its JT8D engines hung close under the wings. Another unusual feature was that it had the same body cross-section as the big 707 and 727, and almost the same cockpit. This offered important reductions in cost and training, and increased passenger appeal with a capacious interior, but there was only one customer: Lufthansa of Germany. Boeing dared not go ahead with such a lone foreign order; but not to go ahead meant no 737 and a monopoly for Douglas. Lufthansa held a board meeting on 19 February 1965 to decide on the proposed 737. An hour before the meeting the airline took a call from Douglas: Eastern was buying the DC-9. This looked like curtains for the 737, but Lufthansa got a personal assurance from Boeing the 737 would go ahead. It was the biggest gamble ever. Lufthansa agreed to buy 21 of the new jets; if this remained the only order Boeing would lose maybe a billion dollars. It all depended on mighty United whether the 737 would really go or not. United wanted extra seats urgently, and did not want to wait for the 737, but Boeing offered early delivery of 727s on good terms to tide the airline over; then, in 1969 or 1970, it would take them back if traffic had not grown to fill them. On 5 April 1965 United signed for 40 737s, 26 727s and 25 extra 727s on lease, plus options on another 30 737s and nine 727s.

To say Boeing was relieved would be an understatement. The 737 flew on 9 April 1967, by which time this stumpy-looking model had become the 737-100. Most customers opted for the stretched -200, seating up to 130. Boeing kept cranking in improvements, such as a unique 'gravel kit' to enable the 737 to operate from rough unpaved airstrips, extra fuel for longer range with automatic 'performance reserve' in the event of engine failure, long-range executive models more spacious even than the competiting DC-9s, and quick-change convertible passenger/freight versions. Gradually the 737 drew level with the DC-9 in the rate at which it was selling, and then it drew ahead – although because of the late start and loss of massive US operators, it trailed in deliveries. But by the late 1970s it was clear that the 737 had become the fastest-selling transport of all, the last to slow in a recession, the first to pick up after-

Above: Much larger than Mirage III fighters, the IIIV was developed from the Balzac-001 as a possible Mach-2 VTOL. Here the eight RB.162 lift jets can be seen.

wards, and the preferred choice for a remarkable number of customers. In the 1980s it was by rights becoming obsolescent, but then British Airways bought 28 and Lufthansa another 32, while for the British package holiday market it was almost the standard vehicle. By this time sales had neared 1,200, for almost 120 customers, and overtaking the DC-9 seemed merely a matter of time, especially with the 737-300 powered by quieter and more efficient CFM56 engines.

Had Boeing gone to borrow money at the start of the 737 story, the lender would have asked "How many have you sold?" and refused to lend. It would have been useless for Boeing to say it could sell 100 or more – let alone 1,200. One company which did try to borrow was Pratt & Whitney of Montreal, Canada. It asked the Canadian government or the government of Quebec, to help it develop a small turboprop engine, the PT6, to take over from the famous Wasp piston engine in the 500–600 hp class. In 1957 there were no customers, though the company said it had been assured of immediate total sales of 40. This seemed unimpressive, so money was refused. In 1958 the company, a subsidiary of a large conglomerate called United Technologies, boldly went ahead on its own and with its own money built and ran a prototype PT6 in 1959. Then it did get support from the Department of Industry, which has since paid for about 30 per cent of the total development cost through the 1960s of $45 million. Sales were agonisingly slow to materialise. In 1960 a PT6 was sold to power the Hiller Ten-99 helicopter, but this never went into production. Others went into Lockheed helicopters, which again did not go into production. In December 1963 de Havilland of Canada put a PT6 turboprop into one of their popular Beaver light transports – one of the so-called 'Bush aircraft' used for utility transport on wheels, floats and skis – and this sold in ones and twos. With painful slowness the Canadian turboprop began to find customers, until the trickle became a stream, and the stream a flood. Today, the Montreal plant employs 6,300 people and is fast coming up to its 25,000th engine – each priced, incidentally, at more than six times the 1960 figure. Aviation is a great business to be in, provided you do not lose your nerve during the first dozen years when everything looks hopeless.

A few thousand of the PT6 engines have gone into luxurious twins (twin-engined aircraft) used by businessmen. Most executive aircraft are owned by large corporations, but some are the personal property of outstanding individuals; for example, several showbiz stars fly their own helicopters, while Herbert von Karajan, the orchestral conductor, flies his own twin-jet. Business flying was a matter of fewer than 100 aircraft prior to World War II, although even then some of the big oil and manufacturing companies had their own airliners of the same types as used by the scheduled operators. After World War II there was a boom in converted attack bombers, which in an era of cheap fuel were attractive in low first cost, high speed and long range, carrying just three or four in great comfort (though with the basic vibration and noise problems of piston engines). In 1956 the USAF issued a UCX specification (Utility Cargo Experimental) which was met by the Lockheed JetStar, and a UTX (Utility Trainer Experimental) which was met by the North American Sabreliner. Both were attractive high-speed jets, the former with four engines each of about 3,000 lb (1,361 kg) thrust, seating around 11, and the latter with just two such engines and seating about six.

From these grew the family of bizjets

(business jets) which first became important in the 1960s and today number some 5,000 aircraft, each priced at from £2 million to £4 million. If a genuine case can be made for such equipment it becomes a tax-deductible business expense and the saving in time, nervous energy and missed connections in comparison with using public transport is enormous. Not only can highly paid staff get on with their work whilst in transit – and often can hold airborne meetings they could never have arranged on the ground – but they can travel direct to their destination, accompanied if necessay by heavy or bulky items such as piles of engineering drawings or computer printout, spare parts, delicate display models and finished products. Such travel is especially important in less-developed countries of large geographical extent. Even in the USA more than 12,800 of the 13,500 licensed airfields have no regular air service; even where there is an airline it may mean a wait of from hours to days to get to the final destination.

There are something like 35,000 other registered business aircraft, most of which are piston-engined. Twin-turboprops numbered around 4,000 in 1984, and were the fastest-growing sector, largely because they burn much less fuel than equivalent jets, and today fuel is no longer cheap. In 1960 the cost of fuel was typically half the cost of flight-crew and maintenance salaries, whereas today it is typically four times greater. In any case, as noted in the final chapter, business turboprops today can have flight performance hardly any different from the jets. The first business jets, which came on the market around 1960, had turbojet engines which were generally tough and reliable but burned fuel rapidly and were extremely noisy. Thanks to the pressures of competition, environmentalists and soaring fuel prices, the picture has changed completely. Hardly a single modern bizjet is being produced with anything but quiet, fuel-efficient turbofan engines. For example, the first DH.125 bizjets of 1962–63 had Viper 520 turbojets rated at 3,000 lb (1,361 kg) thrust; very noisy, they burned fuel on takeoff at the rate of just over 3,000 lb (1,361 kg) per hour, and the aircraft themselves had a maximum weight of 20,000 lb (9,072 kg) (19,000 lb/8,618 kg, for the prototypes) and carried up to six passengers about 1,200 miles (1,930 km) at around 440 mph

Below: Three of the first F-4B Phantom IIs are seen serving with Detachment A of US Navy fighter squadron VF-101 in 1961, the first unit to fly the fighter.

(708 km/h). Today's British Aerospace 125-800 has TFE731 turbofans, each rated at 4,300 lb (1,950 kg) takeoff thrust, but which put out less than one-hundredth as much takeoff noise and burn fuel at only 2,000 lb (907 kg) per hour, yet the aircraft can weigh up to 27,400 lb (12,430 kg) and carry 14 passengers 3,530 miles (5,680 km) with full reserves and allowances, cruising at 461 mph (741 km/h). In a hurry the Dash-800 can make 533 mph (858 km/h). This shows the fantastic progress that has been made in bizjets since 1960, even though the 125 of 1962 looks almost the same as the 125 of 1984.

Of course, the Jet Set that gives this chapter its title do not normally have access to their own jet aircraft, but as a group typify the way air travel has become relatively cheaper even when using public transport. Admittedly the Jet Set are by definition well-heeled, and have sufficient time to be seen in all the most fashionable places, but the impact of the jet on society has been to allow an increasing proportion of the population to go to distant places without taking up any significant time in getting there. Thus, the Jet Set can go to the Edinburgh Festival, nip across to an art exhibition in New York, back to a party in the south of France and then take part in a yacht race at Sydney, Australia. Some manage to fit this kind of social calendar in with a full-time job. As for the general population, air travel has become so simple, reliable and relatively cheap that great distances are no longer a barrier. For a European to have a daughter in the USA or grandchildren in Australia does not mean they cannot be visited; it is instead a reason for making the journey. Perhaps the biggest single growth area in all human travel has been the packaged holiday, which today is responsible for some 30 million air journeys annually. Even this is only a small proportion of the annual total of flights by regular airlines. Today a billion passengers a year travel in, or from, almost every country in the world (the exceptions are tiny principalities too small to have an airport).

In developing countries, such as most of those in Africa and Latin America, there are often few surface communications links and so almost everything and everyone travels by air. This can even extend to cattle going to market or children to a distant boarding school. It is not uncommon for air transport to be provided as a kind of social service by an airline run by the air force, or even as part of the air force. Until the 1960s almost all the large transports operated in 'outback' regions continued to be piston-engined, either wartime ex-military machines or piston-engined airliners prematurely retired because of jet competition. Examples of the former naturally included the DC-3 family, hundreds of which are still in use all over the world, while in South America the C-46 Commando, a much larger twin-engined machine, was also popular. By far the most important of the former airliners were the Lockheed Constellation family and the various types of DC-6 and DC-7. During the 1960s thousands of tired aircraft built in the 1940s were at last allowed to rot, or were scrapped, and much smaller numbers of expensive new replacements were purchased.

By the end of the decade it was becoming apparent to a growing body of enthusiasts all over the world that the attrition of old aircraft had removed from the scene the last example of many famous types of aircraft. In 1945 surplus aircraft were a major problem. On a single field in Germany 148 Martin B-26 Marauder bombers, most of them almost new, were blown up by strapping an explosive charge under the belly of each. This reduced them to pieces that could be carried away in trucks. At Dumbarton, Scotland, brand new Sunderland flying boats were taken off the production line, launched into the water, stuffed full of thousands of equally new items such as engine parts, radio sets and ammunition, and then towed out to sea and sunk. At Thornhill, Southern Rhodesia, the author helped break open 225 giant crates, take out Fairchild Cornell trainers that had never flown, and laboriously break them up and burn the combustible parts. In the Arizona desert Davis-Monthan AFB (Air Force Base) became the site of MASDC (pronounced maz-dick), the US Military Aircraft Storage and Disposition Center. Lines of unwanted aircraft stretched for miles, perfectly preserved in the dry air: fighters, helicopters, giant bombers and transports, all parked wingtip to wingtip. In recent years, the population has varied between 5,000 and 10,000, but changes constantly.

In the late 1940s nobody bothered about preserving aircraft. The unique DC-1 was scrapped in Spain, the Lancaster that had flown the greatest number of combat missions was scrapped in Britain, and countless other famous and unique aircraft likewise vanished. But by the 1960s there was a fast growing army of enthusiastic researchers, chroniclers and preservers. Aviation archaeology became a major enterprise, one aspect of which is searching for interesting relics at crash sites. Surviving examples of old aircraft were lovingly

Right: A trio of Lockheed TF-104G Starfighter trainers, 220 of which were built in California. Most went to the West German Luftwaffe.

Right: Powered by a British Avon turbojet, the Ryan X-13 Vertijet hangs from a platform like a moth, with VTOL (vertical takeoff and landing).

restored, the 'Confederate Air Force' organised the world's biggest collection of actually flying World War II aircraft in Texas, and aviation museums sprang up all over the world. It was the start of a new era in the second half of the twentieth century that affords millions of enthusiasts a personal involvement in aviation in such fields as homebuilts, microlights, hang gliders and, not least, the preservation and restoration of historic aircraft.

At the same time, the mainstream business of manufacturing became ever more remote from the ordinary person. Each new major programme called for a high-risk investment of typically $1 billion (about £600 million), which daunted even the giants of the US industry or the conglomerates that resulted from the mergers in Europe. In both commercial transports and combat aircraft, one answer was for two or more nations to collaborate. The Communist Warsaw Pact (WP), formed in 1955, has remained dominated by the Soviet Union, and true collaborative projects have been rare. In general, combat aircraft for all WP forces are produced in the Soviet Union, while GA (general aviation) aircraft for all WP nations, including the Soviet Union, are produced in Poland. In NATO, however, true collaboration has been attempted in many programmes, and where it has collapsed it has been because of freedom of choice.

There are many ways of arranging collaboration. In the search for a long-range ocean patrol and ASW aircraft to replace the P2V (from 1962 called the P-2) Neptune, NATO held a competition to pick a design, won by Breguet of France with the Br.1150 Atlantic. This was then shared out between factories in France, Belgium, the Netherlands, West Germany and Italy; Britain shared in the 6,100-hp Rolls-Royce Tyne engines and 16-ft (4.88-m) propellers, and the USA supplied many of the advanced sensors and other avionics (aviation electronics). The first Atlantic flew in October 1961 and 87 were delivered.

In the same year production started in many European factories of the extremely fast (Mach 2.2/1,450-mph/2,335-km/h) Lockheed F-104G Starfighter; this was adopted by West Germany, Italy, the Netherlands, Belgium, Norway, Denmark, Canada and Japan, as a multi-role fighter-bomber and reconnaissance machine. First flown in the USA in 1954, the F-104 was not wholly successful, but the F-104G was a more advanced model notable for its self-contained INS (inertial navigation system) with which it became one of the first aircraft able to navigate accurately in any weather and without any outside help. In this case it was a straightforward instance of the USA selling a product to many air forces, which then built their aircraft in a giant multinational programme. On the other hand, when in 1965 France and Britain wished to develop an advanced trainer and light attack aircraft they joined forces from the start to create the Jaguar, forming a two-nation company called SEPECAT in 1966 to manage the project and sharing the manufacturing 50/50, with no duplication of manufacture but with an assembly line in each country. This was soon found to be the best method although, as described in the next chapter, it was not followed with Concorde.

Early in the 1960s the whole of NATO went mad over the idea of jet V/STOL (vertical or short takeoff and landing) aircraft. To a small degree this was because the advancing technology of gas-turbine engines made it possible to build a jet flying machine that could rise straight up, as had been demonstrated by such strange devices as the Rolls-Royce 'Flying Bedstead' and SNECMA 'Flying Atar' of the 1950s, and the even stranger delta-winged Ryan X-13 which at rest hung on a trapeze like a giant moth. To a greater extent it stemmed from the recognition – apparently forgotten today – that NATO airpower is concentrated on a handful of immovable airfields, all of which are constantly targeted by numerous Soviet missiles which could wipe out that airpower without any warning.

A third motivating factor was provided by the engine companies which thought they could see good business. This was especially the case with Rolls-Royce, which strongly pushed the official British case for having batteries of anything from eight to 32 specially developed lift jets in each aircraft purely to lift it during takeoff and landing. In flight, the aircraft would accelerate forwards until the wing alone was able to support the weight, whereupon the batteries of lift jets would be shut down and enclosed by doors to give a streamlined shape. The Short S.C.1 first tested this idea

Overleaf: Two Bell AH-1S HueyCobra attack helicopters of the Japan Ground Self-Defence Force. Each armed with eight TOW anti-tank missiles.

JG-
3401
JG-
3402
陸上自衛隊

Above: US Navy No 142259 was the original prototype Phantom II. Known as the F4H-1 (later F-4A) it had a smaller nose than subsequent Phantoms.

in 1957, and by the 1960s was investigating the possibility of such things as making a blind VL (vertical landing) in bad weather. Dassault fitted the same lift jets into a Mirage III, resulting in the Balzac of 1961. By this time almost every major aircraft company in the Western world was frantically trying to come up with the best V/STOL fighter to meet the demands laid down by NATO.

Even British companies could take part, because while new combat aircraft were not to be mentioned in Whitehall, this was for NATO. Hawker had the nerve to put its own money into building a radical little prototype called the P.1127, powered by a totally new kind of engine produced at Bristol which blew its jets through four nozzles, two on each side, which could be swivelled to point downwards to give lift, or rearwards to give thrust. It first hovered a few inches off the ground in September 1960, and a year later was making complete 'transitions'. It could either take off like a normal aeroplane, come to a halt and hover and then go on and make a normal landing, or it could do the opposite and rise vertically, accelerate to the speed of sound and then slow down again and make a vertical landing. But NATO wanted a bigger, faster aircraft, and so Hawker offered the P.1150, with a boosted version of the P.1127's Pegasus engine, and finally the P.1154 with a super-powerful BS.100 engine.

The P.1154 won the contest, but Dassault of France angrily said it would have nothing to do with the British jet, while Rolls-Royce, who wanted to supply eight RB.162 lift jets for each of Dassault's Mirage IIIVs, lobbied furiously to have the French aircraft adopted by the RAF and every other air force in sight. The whole contest collapsed because no air force had actually agreed to buy anything, and by 1965 most had forgotten all about V/STOL, taking their cue from the USAF which for some reason dismissed the idea that anyone might knock out its airfields.

Rather surprisingly the British officials were now once again talking about manned military aircraft, and Hawker eventually got the job of building versions of the P.1154 for both the RAF and Royal Navy. This was a big challenge but the result would certainly have been a world-beater, and in fact very like the aircraft once again being studied in the mid-1980s. But in 1964 Britain elected a Labour government which had campaigned furiously against the British aircraft industry. Numerous figures of astronomic size were produced – such as £250 million, varying up to £1,200 million – to show how much would be saved by scrapping all the British aircraft programmes and buying from the USA. P.1154 was promptly cancelled, along with the jet-lift WG.681 transport, and the Americans were given big orders for Phantom fighters and C-130 transports. To make the deal appear less one-sided, Rolls-Royce were given a contract to redesign the Phantom to take the British engine-builder's Spey engines. This proved a technical and financial mistake of the greatest magnitude, because it resulted not in a better Phantom but in a much worse one, delivered late and for three times the estimated price (costing more, in fact, than the P.1154).

Had it been left alone the Phantom would have been in many ways a fine aircraft, because it was the top all-round combat aircraft of the 1960s by a wide margin. First flown by McDonnell Aircraft in May 1958, it went into production for the US Navy as a singularly big and powerful two-seat all-weather interceptor, which also had considerable attack capability. Powered by two of the very advanced General Electric J79 engines of the kind also used in the F-104 and B-58, it soon proved itself the fastest, fastest-climbing and highest-flying normal aeroplane in the world (the only exceptions were the special research aircraft carried aloft slung under 'mother aircraft' and quite impractical for normal use). Early Phantoms notched up almost every world record available, and proved to be far in advance of all the competition in almost all respects. Particularly noteworthy was the powerful radar, the carriage of four large AAMs recessed under the fuselage (as well as various other weapons on wing pylons), and the great fuel capacity for long range and endurance, backed up by inflight-refuelling capability. In an almost unprecedented move in 1962 the USAF selected the Navy fighter as an interim type for Tactical Air Command. Problems with its

Right: Rushing a casualty to a Bell UH-1 Iroquois helicopter. Called the 'Huey', it has been built in greater numbers than any other Western aircraft since 1945.

F4H-1

own machines, and unexpected outstanding results with the F-4C 'minimum change' version of the Phantom, combined with increasing involvement with the war in Vietnam to lead to repeated USAF orders, first for the specially equipped F-4D version and finally, in 1967, for the definitive F-4E. The F-4E had more power, more fuel, a new radar and an internal gun, a rapid-fire 20-mm cannon under the nose. This last item had long been requested by pilots in Vietnam who for political reasons had been ordered to close to visual identification range with enemy fighters, by which time they could no longer use their missiles. In 1972, the F-4E was given powerful wing slats to improve its power of manoeuvre. This became the standard export version, and the last 138 were assembled in Japan to bring the total number of Phantoms built to 5,211. This easily exceeds the total for any other Western military aircraft since the F-86 Sabre, except for the H-1 helicopter.

This helicopter deserves special mention, partly because of the enormous numbers built – some 15,000 – but also because of its exceptional development over the years to carry heavier loads over greater distances, and also to perform totally different tasks. Early helicopters were utilitarian vehicles useful only because of their unique ability to hover. Compared with aeroplanes of equal power the helicopter is bound to be very inferior in terms of payload and range, and it also usually has inferior speed, the upper limit being well below 200 mph (322 km/h) except for special research machines. In addition, all helicopters suffer

Below: Certainly the most mysterious aircraft ever created in the Western world, the Lockheed U-2 was originally a stratospheric jet-propelled glider. By 1968 the much bigger and heavier U-2R was in production (here is the R prototype) with a diverse load of reconnaissance sensors.

from the arduous operating conditions of the rotating and oscillating parts, many of which are single metal linkages (with none of the 'belt and braces' redundancy and duplication usually possible with aeroplanes) whose breakage would cause an immediate crash. Though it would be mistaken to regard even early helicopters as dangerous, the fact their insurance premiums are several times greater than those for aeroplanes of equal value speaks for itself.

The coming of the gas-turbine turboshaft engine in the 1950s dramatically improved the helicopter scene. The first turboshaft helicopter in production was the little French Alouette, of which well over 3,000 were built from 1955, with manufacture under licence still continuing in Romania. Turbine engines greatly reduced the empty weight of helicopters whilst simultaneously increasing the installed horsepower, eliminating the need for engine cooling (often a problem with slow or hovering machines), reducing noise and vibration, enabling safer fuels to be used and almost eliminating the previous major problem of engine failures. In the Soviet Union the design bureau headed by Mikhail Mil quickly became a world leader with turbine helicopters and in 1957 flew the first Mi-6, a gigantic machine which even to this day has no equal in any other country. More than 800 were built through the 1960s, together with special crane versions, the Mi-10 and -10K, which straddle their load or suspend it from a cable. Another Mil helicopter, the Mi-8, appeared in the 30-seat class in 1961, and since then over 8,000 of many versions of this have been built. But even this number is barely half the total for the H-1 family.

Bell Helicopter was picked in 1955 to build a new turbine-engined utility helicopter for the US Army, and the first example flew on 22 October 1956. It was powered by an Avco Lycoming T53 engine of 700 hp, had five seats and weighed 5,800 lb (2,630 kg) fully loaded. By 1959 slightly improved models were in production, and through the 1960s the new helicopters, popularly called Hueys from their original designation of HU-1, grew in capability as the power of the T53 gradually increased to 1,100 hp. The war in Vietnam multiplied the demand, and thousands were used and expended in that conflict as the standard troop carrier and utility machine. Many were fitted with various kinds of armament, and on 7 September 1965 Bell flew the first AH-1 HueyCobra, a completely redesigned model with a slim fuselage like a fighter seating a gunner in front, working a mass of weapons, and the pilot behind at a higher level. The Snake, as it became called, was instantly popular for jungle warfare, and showed its ability to stay flying even when severely damaged by intense fire from hostile troops. The Cobras were some of the first uprated versions with a Canadian Pratt & Whitney PT6 engine with two power sections putting out 900 hp each and giving twin-engine security. Other Hueys were fitted with Lycoming T55 engines of up to 2,930 hp. By the 1980s development had even progressed to the Model 214ST SuperTransport which can carry 18 passengers or a cargo payload 50 per cent greater than the first Huey's total laden weight! The 214ST has two General Electric engines of up to 1,725 hp each.

Bell is also the builder of the JetRanger, the most popular of all light executive helicopters and also widely used by military customers. Again there are many versions, some of which seem to be constantly on TV screens being flown by cops or 'baddies', and most have either an Allison 250 (T63) engine of from 300 to 650 hp, or a pair of such engines. During the 1960s the number of such 'jet helicopters' in use, of all makes, increased from roughly 180 to more than 20,000!

On the other hand the number of large jet bombers fell in almost the same proportion. At the start of the decade thousands were in use and some nations were busy

Below: A General Dynamics large-winged F-111A attack aircraft on a combat mission from Takhli in 1972 with cluster bombs.

Above: The pilot and RSO (recon systems officer) of the Lockheed SR-71 wear astronaut suits and plan each flight like a space mission.

with next-generation successors. Sukhoi in the Soviet Union was about to fly a large bomber able to cruise at more than Mach 2, while in the USA since the mid-1950s North American had been working on one of the most powerful and challenging aircraft of all time, the XB-70 Valkyrie. Powered by six 30,000-lb (13,608-kg) thrust General Electric J93 turbojets, it was a gigantic canard (tail-first) delta made of stainless steel. Its giant inlets and nozzles were fully variable to match the airflow to the very contrasting conditions experienced at Mach numbers up to 3.2. At this speed the XB-70 could cruise for over three hours with the outer wings hinged downwards to ride in a favourable way on a high-pressure area caused by the supersonic shockwaves. The first XB-70, painted white, flew on 21 September 1964, but the entire programme was eventually cancelled. One prototype crashed after it had been struck by an F-104 which had been flying in formation with it.

In the Soviet Union the MiG-25, called Foxbat by NATO, was specially designed to intercept the B-70, and it was continued after the bomber's cancellation. In April 1965 a prototype set a world record speed of 1,442 mph (2,320 km/h) over a 621-mile (1,000-km) circuit with a 2-tonne (2,000-kg) payload, and later another MiG-25 flew round a 311-mile (500-km) circuit at 1,853 mph (2,983 km/h). Naturally this large and powerful machine caused a scare in Western military circles, and quite large numbers have since gone into service with several countries as all-weather interceptors and uninterceptable high-flying reconnaissance aircraft. But in the latter role even the MiG-25 is outclassed by an aircraft built by Lockheed for the USAF.

Ever since 1943, when Lockheed's design czar C. L. 'Kelly' Johnson had created it to build the secret new XP-80 Shooting Star jet fighter, Lockheed had operated a very small and highly secure 'plant within a plant' known as 'The Skunk Works'. In 1954 Johnson got the go-ahead on an amazing new aircraft to fly higher than any other in history and make clandestine reconnaissance flights over the Soviet Union and other potentially hostile countries. Powered at first by a J57 turbojet and later by the bigger J75, it resembled a sailplane with graceful long-span wings and landing gears under the fuselage, the small outrigger wheels being dropped after takeoff. It was given the utility designation U-2 as a cover, and was announced as a research aircraft for studying the upper atmosphere. But on 1 May 1960 the true purpose – which had been all too obvious to the Soviet Union since 1956 – was suddenly making headlines in every newspaper on Earth. A U-2, painted black and without markings, and operated by the US Central Intelligence Agency, was shot down by SAMs whilst it was photographing Soviet ICBM (intercontinental ballistic missile) developments. The pilot, F. G. Powers, survived and was put on trial; two years later he was exchanged for a Russian spy who had been caught by the USA.

Subsequently there were many bigger and even more capable U-2s, and Lockheed's Skunk Works began work on a totally new successor in 1957. This time mere height was not enough; the highest possible speed was also called for. Johnson's extremely talented team created an aircraft made almost entirely of titanium and its alloys, with a very long slim body projecting far ahead of a stumpy delta wing on which were mounted two giant Pratt & Whitney J58 bypass jet engines. Everything about this aircraft, the Lockheed A-12, was reaching far into the unknown and the first

example flew on 26 April 1962 in complete secrecy. Among the production aircraft was a small group of interceptor versions, the YF-12A, first flown in 1964. By 1965 production had switched to the most important model of this family, the RS-71. The replacement for the A-12, this was longer and heavier and, with much greater fuel capacity, could fly intercontinental missions at greater speeds and heights than any other aircraft, apart from the X-15 which will be described later. Known as Blackbirds because of their colour, these great aircraft are flown by a pilot and reconnaissance systems operator wearing Astronaut space clothing. In 1965 a YF-12A had set a world speed record at 2,070 mph (3,331 km/h), but the RS-71 was even faster and on 28 July 1976 set the current record at 2,193.17 mph (3,529.47 km/h) or Mach 3.32. Not least of the many curious facts about the Blackbirds is that when President Johnson disclosed their existence he announced the A-12 as the A-11 and the RS-71 as the SR-71, and both designations inevitably stuck. This may merely reflect the atmosphere of CIA-style secrecy that has always enveloped these impressive aircraft, but as examples of the design of advanced manned aeroplanes they remain unrivalled even today.

Almost the only part of the Blackbirds that does not have variable geometry (ability to change its shape) is the wing, yet the variable-geometry wing matured technically at just the time they were being designed. The term is today usually taken to mean variable sweep, the wings being pivoted so that they can spread out to large span, with high aspect-ratio, for a short takeoff or for efficient low-speed loiter over extended periods, and then folded back into an acute delta shape for a supersonic dash. At low level the VG (variable-geometry) wing has the further advantage of giving very small span and a high wing-loading, so that an attack run can be made in dense turbulent air without shaking the pilot's eyeballs out. Today there are many so-called attack aircraft with large fixed-shape wings whose alleged supersonic speed capability is compromised by the inability of the pilot to stand the rough ride.

VG wings were studied in Nazi Germany, and one prototype, the Messerschmitt P.1101, was taken to the USA and turned into the Bell X-5, the first such aircraft to fly (on 20 June 1951). The idea was then tried in the Grumman XF10F Jaguar carrier-based fighter of 1953, but this was so complicated and clumsy that little more was done for six years. Sir Barnes Wallis at Weybridge then perfected a neat two-pivot scheme, but instead of building it the British government ordered that it should be handed over to NASA in the USA (just as had been the case with the cancelled British Miles M.52 supersonic research aircraft of 1946). NASA and the USAF then came up with the outline scheme for a TFX (tactical fighter experimental) embodying not only variable sweep but also augmented turbofan engines and many other advanced features. The way was open to a massive increase in all-round capability for a multi-role combat aircraft, with more than doubled range and endurance, much heavier loads of conventional weapons, advanced radar and AAMs and a dash speed of well over Mach 2. TFX was the subject of a giant industry-wide competi-

Left: One of the first releases from the NB-52A mother ship of the first of three X-15 hypersonic research aircraft in 1959.

tion from 1960, which after unprecedented study and refinement was won by General Dynamics in 1962.

The new fighter was designated the F-111. It was planned as the greatest combat aircraft ever, to replace all the fighters, tactical attack and long-range interdiction, air superiority, long-range all-weather interception, nuclear strike and reconnaissance aircraft of both the USAF and US Navy. Thousands were expected to be built, hundreds of them for export, but there were problems. Some lay in the specification; the range and supersonic endurance were pitched at such challenging levels that the aircraft dimensions were enormous, and it became too heavy and unwieldy to be a good fighter. The attempt to meet Air Force and Navy needs in one design proved to be an impossible objective, and after increasingly frustrating problems the Navy finally pulled out in 1968, cancelling their F-111B version. There were also political problems; both Congress and the Pentagon were bitterly divided over the programme and, while some said the government had "bought the second-best airplane at the highest price" others said the aircraft was in any case 'a lemon' (slang for incapable and useless).

The first F-111 flew on 24 December 1964, and soon ran into a series of further problems. The engines failed to work to such a degree they were called 'a hazard to safe flight'; the inlet ducts had to be redesigned; the drag was much worse than predicted; the weight was up by more than 50 per cent and far more fuel had to be squeezed in to try to get somewhere near the desired range. A USAF general had said "I'm not going to accept any goddamned 75,000-pound airplane", yet it came out with a gross weight of 100,000 lb (45,360 kg), and grew from there. Entering USAF service in 1967, it proved so useless as an air-combat fighter that it carried no air-combat weapons and only Sidewinder AAMs for self-defence. Six went to Vietnam and three were lost in as many weeks. Another crashed in the USA and it was found that a tailplane actuator was faulty. Then wings started coming off, and every F-111 had to be expensively proof-tested. A batch bought by Australia was put into storage in Texas, undeliverable, and finally got to Brisbane just ten years late, in 1974.

All this obscured the basic fact that the F-111 was in its day the world's best long-range attack aircraft. If it had not been thought of as a fighter or forced on the Navy, its development would have been quicker and more successful, and it would not have become the favourite butt of criticism for media all over the world. Almost the only place where the F-111 was for a time lauded as a great wonder-plane was Britain, because here the government had abruptly cancelled its own very successful TSR.2 in April 1965, and said it would save money by buying the F-111 (as it had 'saved money' a year earlier by replacing the P.1154 by the Phantom). In the event plans for a British F-111 were quietly shelved and a big cancellation payment was made to General Dynamics. The RAF F-111 was cancelled just as it had matured as a world-beating bomber, and in Vietnam the F-111 demonstrated for the first time a new style of air attack. Using TFR (terrain-following radar) they could thunder down amongst cloud-covered mountains, diving and climbing along the undulating terrain to avoid the enemy radars and occasionally giving the pilot and navigator, sitting side-by-side, glimpses of jagged rock a few feet off the wingtip or under the nose just to keep the adrenalin flowing. At the target they could put down bombs within a 50-foot (15-m) circle without having to see it. As never before, the F-111 crews put their lives into the small electronic 'black boxes' that steered the aircraft unerringly within 100 ft (30 m) or so of the very hard ground. Sadly it was unfashionable to praise the F-111.

The F-111 typified the new age of aviation in which to conquer darkness, fog and guidance problems, avionics has become more important than mere speed. Yet speed still reigned in the 1960s. As mentioned earlier, the North American X-15 was the fastest and highest-flying aircraft ever built. It was planned in 1954 as the next-generation successor to the original series of X-1 high-speed research aircraft which were carried aloft by a 'mother ship' to conserve their limited rocket fuel. The programme was conceived by the NACA, which in 1958 was restyled the US National Aeronautics and Space Administration (NASA), while the funding and detailed engineering management was provided by the USAF. North American built three X-15s, the first flight under rocket power being made on 17 September 1959 after being dropped by its NB-52 parent. Gradually through the 1960s the X-15s worked their way up the scale of speed and height. Made of titanium, and filled with highly reactive liquids, the X-15s were rather tricky to handle and their landing gears comprised a conventional twin-wheel nose gear and a pair of steel skids under the tail. The greatest height reached was 354,200 ft (108 km), and the highest speed a stunning 4,534 mph (7,297 km/h), or Mach 6.72, on 3 October 1967. The pilot on this occasion was a NASA civilian, William J. Knight.

Today the last X-15 is in a museum, and not only is there no aeroplane even half as fast but there is no plan to build one. Yet it would not be right to regard the 1960s as the end of the quest for greater speed and height in flight through the atmosphere. The horizons today are even wider, but inflation puts sharp limits on what can be afforded.

The US Marine Corps uses the KC-130 Hercules in the combined tanker/transport role. Here a KC-130R of squadron VMGR-252 replenishes two F-4J Phantoms.

9. TAILS OF YESTERYEAR

In aviation it seems that one often has to run furiously to stay in the same place. Whereas in the 1930s, 1940s and 1950s technical progress had galloped ahead at a fantastic rate, by the 1960s one often seemed to be standing still. This is despite the fact that, compared with earlier decades, the research and development teams of top engineers were bigger than ever before. For example in the 1950s the number of design engineers in the British aircraft industry was on average around 3,100, and they made fast progress with over 180 aircraft projects. By the 1970s the equivalent number of top technical staff was over 7,000, yet they had great difficulty in getting anywhere with a mere handful of aircraft projects – the three biggest were cancelled in 1964–65, leaving little but two joint programmes with the French, Concorde and Jaguar.

The spectre of inflation galloped ahead much faster than the technology of aviation through the 1970s, putting many good ideas financially out of reach and making a mockery of many carefully considered statements of a few years earlier. In 1967 the assistant managing director of Hawker Siddeley Aviation said "We can sell you a Harrier for around £750,000, depending on how many you want and the standard of equipment"; today £7.5 million would not be enough. In 1972, after acrimonious court hearings concerning the F-14 Tomcat, a top US Senator said "Never again will the US government buy a $20 million fighter"; how right he was, because today they are never that cheap!

Superimposed on general inflation was another factor that from 1973 totally altered all the budgets and cost analyses of all users of aircraft, and the public air carriers in particular. This was triggered off by a so-called 'fuel crisis'. There was no real crisis at all, in the sense that the Earth's petroleum reserves and extraction and refining capacities were unchanged. What happened was that in October 1973 another war broke out between the Arab countries and Israel, and the Arab oil producers chose to put a ban on their oil exports, later changed to resumption of supplies at a reduced rate. They had, perhaps wisely, decided they might be short-sighted in selling their birthright, and their only big exportable asset, at the most rapid possible rate and for the lowest price, of $1.5 per barrel. Within a year the price of a barrel had climbed to $30.

This incredible and unforeseen rise in fuel price was paralleled in the other oil-producing countries, and for a while private

Right: Ever since spring 1982 air and naval staffs have been studying the brilliant results achieved in both air combat and surface attack by the Royal Navy's Sea Harriers.

Below: Airline operators, and indeed air forces, have only gradually realised that the British Aerospace 146 has not only hit it exactly right for a very large market sector but is also a superbly engineered aircraft able to replace anything from a DC-3 to a DC-9.

000
ROYAL NAVY
003
001
ROYAL NAVY
N

Above: Created jointly by McDonnell Douglas at St Louis and British Aerospace at Kingston, the Harrier II is in production for the US Marines and RAF.

motorists and other users of petroleum products carefully husbanded their supplies and cut down consumption. Airline pilots cruised at Mach 0. 73 or so instead of Mach 0.84, and took great care to minimise dead legs – between stacks and joining the glide-path, for example – and kept the aircraft in a low-drag condition for as long as possible before finally lowering the flaps in stages to the landing setting and dropping the gears just before coming over the threshold. In 1974 the author returned from Tehran to London flying most of the way in a KLM DC-8, and on 378 occasions the DC-8 crew took particular actions to conserve fuel. They reckoned they were burning 26 per cent less fuel than two years previously, on the same airline sectors and with similar loads. For a time the need to save fuel was manifest in all walks of life, and even today there is often a fuel-price clause written into a passenger's contract on booking a package-tour holiday paid in advance. In fact no meteoric rise in fuel price is now likely, and for several years the trend has been if anything downwards, gradual rises to the consumer being caused by national taxes.

The author's outward trip to Tehran had been made at Mach numbers of 0.92 from London to the Black Sea and 2.05 thereafter. The vehicle was Concorde of course, and no transport vehicle so well demonstrates the ability of the engineer to solve the technical problems but not the political ones. Britain had set up a Supersonic Transport Aircraft Committee in 1956, which eventually recommended choosing a cruising Mach number of 2.2, at which speed the main structure could still be of aluminium alloy. Discussions with the French led to a joint 50/50 programme for Concorde launched in November 1962. At first France had strongly wanted to build a lightweight Super Caravelle for short routes, but after the world's airlines had said the British were right to insist on transatlantic range they agreed to go along with a much heavier long-range machine to suit the market demands. Bristol Siddeley (bought by Rolls-Royce in 1966) developed a new version of its very powerful and efficient Olympus turbojet, the Mk 593, specially for the Concorde. SNECMA, the French propulsion partner, produced an excellent reheat jetpipe, thrust reverser and variable nozzle with special provisions for reducing noise. Airframe design and production was shared equally between BAC and Sud-Aviation, the latter becoming Aérospatiale in 1970 after swallowing groups which had roots going back to no fewer than 37 pioneer aviation companies. In fact the Concorde programme was run

but effectively thrust downwards rather than upwards at takeoff or on landing, just when the aircraft needs extra lift. Thus at maximum weight Concorde has to accelerate to over 230 mph (370 km/h) before lifting from the runway, and a fundamental feature of the design is that the lift at high AOA (angle of attack) is increased by the formation of a giant writhing vortex from the sharp leading edge of the wing, extending back across the wing and dying out astern. In many atmospheric conditions this gives rise to a livid white plume of condensation.

The development problems were not unduly severe, but for various reasons the project ran late and suffered so badly from inflation, multiplied by the delays, that the original cost estimates became a sick joke, the final bill exceeding £1,000 million. The first prototype flew (at Toulouse) on 2 March 1969. There were two such prototypes, followed by two larger and heavier pre-production aircraft, followed by 16 production Concordes, which were much longer and heavier still. Even after delivery small improvements continued to be made, and though seemingly trivial these had a significant effect because in Concorde operations the costs and the revenue are both very large figures, and the mass of fuel is much greater than the mass of passengers, and even small changes to reduce drag and cut fuel consumption have a large effect on economics. Scheduled services began on 21 January 1976, but at this date political barriers were keeping the Concorde out of its most important destinations, notably New York, and the only place that

Below: This test rig looks crude, but it should lead to vectored-thrust engines with PCB (plenum-chamber burning), of the type needed for supersonic V/STOLs.

by a committee of directors from government and industry in both countries, with production lines at both Bristol and Toulouse. It proved an unhappy arrangement which will not be repeated.

From the start the Concorde was an aesthetically beautiful machine with a slender needle-nosed fuselage, so-called ogival delta wing of small span but colossal chord, with two twin-engine boxes under the wing outboard of the landing gears, with fully variable inlets and nozzles under computer control. A feature which caught the public's attention was the 'droop snoot' nose, which in low-speed flight can be reprofiled by retracting a streamlined vizor to give better forward view, and for the landing can be bodily pivoted down to give a view of the runway even in the nose-high landing attitude. Perhaps surprisingly it was decided not to use a foreplane or tailplane but fit the wing with powerful elevons, which are fine for control in flight

would have the British Airways service was Bahrein! It had been planned to extend this route to Singapore and Australia, but – again for purely political reasons – this never proved possible. Eventually services to New York and Washington were permitted, British Airways Concordes at last serving New York from 12 February 1978, cutting the time from Europe to less than $3\frac{1}{2}$ hours.

The violent and prolonged opposition to Concorde, at New York in particular, was an interesting social phenomenon. Even in Britain Concorde was widely presented in the media as a kind of anti-social monster, and an Anti-Concorde Project was organised on a nationwide scale with extensive funds and gaining wide publicity far beyond anything the harrassed planemakers could muster. Many totally false arguments were given worldwide credence, a typical one being that Concorde would somehow eat up the ozone in the stratosphere and expose the human race to death from cosmic radiation. This overlooked the fact that 16 Concordes is a small number compared with the thousands of military supersonic aircraft, and it also overlooked the excellent state of health of the Concorde flight and cabin staff who spend their entire working lives in the supposedly lethal stratosphere, prey to incoming harmful radiations. The inevitable conclusion is that any highly visible technical advance, no matter how far it may be dedicated to the public good and to shrinking journey times and thus bringing people together, will in future become the target of frenzied attacks aimed at destroying the project. It is reminiscent of the earliest days of steamships, steam railways and motor cars, when arguments which today appear laughable were seriously believed by most of the population as

Above and below: First flown in 1954 the Lockheed C-130 Hercules has become the standard military airlifter of most of the non-Communist countries, and despite its age is still in full production. Above, one of the C-130Hs supplied to the Royal Jordanian AF. Below, a standard USAF C-130H of Military Airlift Command on a supply run to the new airport at Grenada in 1984.

reasons for preventing the adoption of such 'newfangled' ideas. Equally, the people who campaigned against Concorde in the 1970s might find it difficult to understand how their descendents a century hence will regard Concorde as a cherished bit of ancient history, whilst at the same time fulminating violently against whatever happens to be new in the late twenty-first century.

Ironically, whilst Concorde itself has performed with outstanding reliability, despite its high level of technical complexity with many extremely advanced systems that are absent from subsonic transports, the operating profitability of the aircraft has been destroyed by the 20-fold increase in the price of fuel since the project was conceived. This was the only thing that could seriously damage the SST, because fuel is by far the biggest item in SST operating costs, unlike other airliners. There are fundamental reasons for this which no amount of technical development can alter. This makes the SST as a class less attractive than in the pre-1973 era of cheap fuel, and it is a major factor that probably was not anticipated by the US Congress when it terminated support for the planned rival American SST – the very big Mach-3 Boeing 2707-300 – back in 1971.

This left only the Soviet Union with an active rival SST project, the Tu-144, and this in fact got into the air ahead of Concorde in 1968. Similar to the Anglo-French machine in configuration, but larger and more powerful, it went through no fewer than three major phases of redesign and even today is not in regular service. The Soviet Union is so large that it saw an SST as a natural future development, and predicted that the large development costs would be speedily recovered in saving the time of millions of long-distance travellers each year. Back in 1963 Aeroflot, the gigantic Soviet civil aviation organisation, calculated it already saved each of its 55 million passengers (one-quarter of today's annual total) 24.9 hours on each journey, and that with a fleet of 75 SSTs the figure would be increased to over 36 hours. There is no reason to doubt the validity of these figures, and the price of fuel in the Soviet Union has not been tied to that in Western markets, but the Tu-144 has been plagued by technical problems which have seriously delayed its development. One of the first redesigned production aircraft crashed at the Paris airshow in 1973, and six months after the start of scheduled services another crashed fatally in the Soviet Union. Next came the third-generation Tu-144D with new engines, but even this had never carried fare-paying passengers by the time this book went to press.

Speed has thus not been a fruitful route for further progress in air transport, and it is extremely unlikely we shall see much advance in speed beyond the level established 30 years ago by the 707 and DC-8 apart from small gains achieved by aerodynamic refinement. On the other hand, radical changes in propulsion in the 1960s paid off in the 1970s with today's widebody airliners which, like the 707 and DC-8 a generation earlier, have arrived at just the right time to enable more people to be moved over all distances at lower cost. By-products of this revolution have been elimination of the previous problems encountered in trying to build really long-range aircraft, and elimination of noise and smoke as major social problems of aviation.

The new propulsion is centred on the HBPR (high bypass ratio) turbofan. This can be regarded as a turbojet driving a propeller encased inside a circular duct. Not least of the paradoxes of aviation is that when British pioneers, including Whittle, proposed such an engine 40 years ago it was ignored. It was loosely thought of as falling somewhere between the very de-

Right: Unlike prototype and pre-production Concordes the production aircraft have a long rear fuselage aft of the wing. This machine was the first French production Concorde.

sirable turbojet and the very desirable turboprop, the former being noisy and burning a lot of fuel but being simple and the obvious choice for high-speed aircraft, and the turboprop being complicated and needing a heavy gearbox and propeller but being economical and the best answer for medium-speed machines. Not until the mid-1960s did engine designers dust off old calculations and update them with higher engine internal pressures and temperatures, and they suddenly found that they could make engines of great power with very low fuel consumption and with dramatically reduced noise. Typically they began thinking in terms of 40,000-lb (18,144-kg) thrust, most of it generated by the giant many-bladed fan in some 8 ft (2.43 m) diameter at the very front of the engine. The outer parts of the fan blades rotate faster than sound, eliminating the need for the inlet guide vanes used on all previous axial-compressor jet engines, which in turn eliminated the need for hot-air de-icing.

The biggest advances of the HBPR turbofan were not only the large thrust but also the unprecedented efficiency, with fuel consumption (for any given thrust) less than half the typical values with a turbojet, and also the enormous reduction in noise. Jet noise varies to an incredible degree with the speed at which the jet comes out of the nozzle. Increasing the speed of a jet from 1,000 mph (1,609 km/h) to 1,100 mph (1,770 km/h) may not seem much but it roughly doubles the noise level. But the HBPR engine replaced most of the jet by a large, relatively slow-moving jet of cool air from the fan, giving hardly any noise at all. There was left just a little fan noise at the front and the noise from the small hot jet at the rear from the 'core engine' driving the fan. To a rough approximation a 40,000 lb (18,144 kg) HBPR engine promised to generate one-hundredth as much noise as a turbojet rated at 10,000 lb (4,536 kg)!

Developing completely new engines of this size costs hundreds of millions of dollars or pounds and is a severe risk unless a military order can be obtained. The only such order in the Western world was the USAF's so-called CX-HLS (cargo experimental, heavy logistic system) requirement of 1964. The USAF had almost bought a giant airlift transport in the 1950s, the Douglas C-132, and the only power then conceivable was the turboprop, four 15,000-hp T57s being chosen. The C-132 never flew, but by 1961 the maker of the T57, Pratt & Whitney, had discovered the enormous potential of the HBPR turbofan and, with the USAF, came up with exciting possibilities. It is remarkable that Pratt & Whitney did not win the contract for the

Below: By early 1985 Boeing had sold well over 600 of the biggest and most powerful jetliner, the Jumbo Jet. Northwest Orient, of Minneapolis/St Paul has 30.

Below: An M-1 Abrams main battle tank, at 60 tons, is no problem for the USAF Military Airlift Command's C-5A Galaxy; the monster can carry two tanks.

Advanced
TriStar
N1011

LOCKHEED L-1011
U.S. AIR FORCE
F-16XL

Previous pages: N1011 was the original Lockheed L-1011 TriStar, retained at Palmdale as the company development aircraft.
Page 203 top: Lockheed-Georgia, delivered 284 C-141 StarLifters to the USAF. Then it took 270 back and, as shown, lengthened them into C-141Bs.

Page 203 bottom: The F-16XL is a total revision of the F-16 Fighting Falcon to a tailless-delta layout.

CX-HLS engine, which went instead to rival General Electric, with its TF39 of 41,000-lb (18,598-kg) thrust. To build the monster aircraft the USAF picked Lockheed-Georgia in October 1965, previously the creator of the C-130 and C-141.

The C-130 could have been mentioned in any of the last three chapters. First flown in August 1954 it was designed in 1951 as a modern military transport, and it marked a complete break with the badly arranged predecessors. Instead of having a narrow sloping floor with access through a small side door it had a large level floor close to the ground, at the same height as typical lorries and army trucks, with access via a vast hinged ramp at the rear up which a large vehicle could be driven. Powered by four Allison T56 turboprops, initially each of 3,750 hp and today of 4,910 hp, it could leap out of a short rough airstrip, climb steeply, handle like a fighter and cruise in pressurised comfort over long ranges at the speed of a Spitfire. The first C-130s entered service in 1956, by which time they had been named Hercules, popularly changed to Herky-Bird. Hundreds were built, some with heated skis for polar service, others with special gear for tanking, refuelling, SAR (search and rescue), launch and control of RPV (remotely piloted vehicle) drones and targets, maritime patrol, mapping and survey, night ground attack in Vietnam with batteries of guns of all calibres and special night vision sensors, and many other duties including retrieval of space capsules in mid-air.

By the 1970s it was clear the C-130 was having an amazingly long life. The maker's advertisements invited military and civil customers to buy this ageless machine 'which keeps acting newer and newer'. Supposed replacements came and went and still the Herky-Bird stayed in production. In the 1970s large sums were spent on two very clever STOL (short takeoff and landing) jets, the Boeing YC-14 and McDonnell Douglas YC-15, specially to replace the C-130. Both remained prototypes while air lines, air forces, navies and other buyers kept ordering C-130s.

One of the other transports that came and went was the much bigger C-141 StarLifter, likewise a product of Lockheed-Georgia Company. Roughly the size and weight of a 707, the C-141 was given a high-mounted wing to bring the giant cargo floor near the ground as in the C-130, and the wing was given less sweep than a 707 because short takeoff was more important than speed. The USAF bought 284, all of which were in service by 1970. They were the transpacific workhorses of the Vietnam war, shuttling westwards with pallets of up to 40 tons of cargo and returning with every care taken to avoid rough air and to make perfectly smooth landings with 80 battle casualties in their litters (stretchers) and 16 more seated, together with all necessary medical equipment and attendants. The only major shortcoming of the C-141 was that its fuselage had the same cross-section as the much smaller C-130, namely 10 ft (3 m) wide and 9 ft (2.7 m) high, and this not only prevented loading the biggest loads but also meant that the fuselage was often physically full long before the payload limit had been reached. In the 1970s the USAF Military Airlift Command decided to stretch its remaining 277 StarLifters by 23 ft 4 in (7.11 m), in effect adding 90 aircraft to the fleet from the viewpoint of carrying capacity without actually adding any. The first of the longer C-141B versions flew in March 1977, and all had been rebuilt by 1982.

From the start Lockheed-Georgia's CX-HLS submission had been of a totally different scale of size with a cargo hold 144 ft 7 in (44 m) long, 19 ft (5.79 m) wide and 13 ft 6 in (4.1 m) high, with a second slightly smaller deck at an upper level. Not only was there a ramp at the rear but the entire nose could swing upwards, exposing the entire interior at both ends. The aircraft was designed to land on 28 soft tyres on any kind of reasonably firm terrain, despite its weight of 800,000 lb (362,870 kg), and the tail stood as high as a nine-storey office block. It was designated the C-5A Galaxy, and the first took the air on 30 June 1968. Lockheed expected to build 115, but like the F-111 the C-5A had been bought by an unfortunate Pentagon process called Total Package Procurement, and severe cost-

Below: Scorned at first as 'a committee-designed aircraft' the Airbus A300 won its place on sheer merit as a worldwide best-seller.

Right: Typical of today's airline engines, the General Electric CF6-50 swallows up to 1,500 lb/s of air to generate a thrust of some 55,000 lb (24,948 kg).

overruns caused a public scandal and termination at the 81st aircraft. Worse, there were annoying mechanical problems and continuing fatigue cracking of the wing – which had been designed under intense pressure by a group of unemployed British engineers made redundant by the many cancelled British programmes – which ultimately forced Lockheed to re-wing the aircraft in the 1980s.

One of the companies that went after the CX-HLS contract was Boeing. This company constantly looks far ahead, and for several years it had been discussing thoughts for the 1970s with civil airlines, notably PanAm. Boeing was certain there would be a demand for a new-generation transport much larger than anything seen previously, and the HBPR engine made it possible. Unlike the C-5A it would have a low wing and be arranged in a very different way, with normal passenger side doors, landing gear tailored to good runways and with the emphasis on speed and range rather than short field capability. By 1966 the basic arrangement was at last settled; instead of a vertical or horizontal figure-8 ('two-tube') body, there would be just one enormous passenger floor all at one level, with seats right into the nose, the flight deck and a few passengers being at an upper level. One of the advantages of this arrangement was that it facilitated the use of standard cargo pallets above the floor in a freight or convertible version. Cabin width was fixed at 20·ft $1\frac{1}{2}$ in (6.13 m), even more spacious than the C-5A, though of course the ceiling height was less. This enabled two wide aisles to be provided, greatly facilitating movement of passengers, cabin staff and trollies, yet with ten-abreast seating (3+4+3) no fewer than 516 passengers could be accommodated. The wing was swept at no less than 37.5°, matched to high cruising speeds approaching 600 mph (966 km/h), and a unique feature was that the giant high-lift flaps on its leading edge were arranged to bend as they were pushed open hydraulically until at full extension they were curved to the best aerodynamic profile. Other unique features were that there were four main gears, each with a four-wheel bogie, and navigation was by triple INS (inertial navigation systems). Another remarkable fact was that the engines were by Pratt & Whitney, loser in the CX-HLS competition, which at its own expense produced a massive HBPR engine designated the JT9D, and initially rated at 43,600 lb (19,780 kg).

The colossal new aircraft, called the 747 but known to the media as the Jumbo Jet, was launched by a PanAm order for 25 placed in May 1966, but it still represented a risk of many times the net worth of Boeing. A gigantic new plant had to be built to manufacture it, and the building, which grew up around the 747 production line in a space hacked out of forest at Everett, north of Seattle, is the world's biggest volume enclosed space. Out of it came the first 747 for its maiden flight on 9 February 1969 (it had been planned for 17 December 1968, anniversary of the Wrights' 1903 flight), and services with PanAm began amid a welter of problems

on 22 January 1970. The problems were centred on the monster engines, which were distorting (in a way called 'ovalizing') and giving particular difficulties in cross-winds. The first PanAm 747s were furnished for only 382 passengers, but the arrival of two Jumbo-loads all at once caused problems at many airports, and some of the growing list of customers specified as many as 490 seats. True to form, BOAC flight crews went on strike for higher wages, and in May 1970 the first British 747s sat on the ground unused. Once the pilots had been bought off, the first service was fixed for 18 April 1971, but then it was the turn of the flight engineers (who had perhaps begun to recognise that they were, as a group, redundant in the modern world). They forced cancellation of the much-delayed first 747 service an hour before takeoff, but the BOAC Jumbos at last started earning, almost a year late, on 25 April 1971.

Subsequently the 747 became the standard mainline vehicle on the busiest and longest trunk routes. It grew in weight to 840,000 lb (381,024 kg), with engines by PWA, GE and Rolls-Royce of up to 56,000 lb (25,402 kg) thrust. The 747F version appeared with a cargo floor, upward-swinging nose and computerised loading; the first 747F delivered to Lufthansa flew the round trip between Frankfurt and New York six times a week without a break for its first three years, each time with a 108-ton (109,735-kg) load. The 747SP (Special Performance) model, first flown in 1975, has a shorter body and bigger tail and, despite flying at reduced gross weight, has ultra-long range and can fly any route in the world. An SP on delivery to South African Airways in 1976 arrived in Cape

Above: Airbus uses a fleet of these Super Guppy freighters to carry American engines, and (as seen here) complete wings from Britain.

Top right: Finland is one of the many countries whose longest airline routes are flown by the McDonnell Douglas DC-10-30 wide-body trijet.

Right: Singapore Airlines has been so satisfied with its A300B fleet, one of which is seen here on test over the Pyrenees, that it is now adding eight of the smaller A310 version.

DC-10

SINGAPORE AIRLINES

Above: The Airbus controls are surprisingly compact. The pilot demonstrates the neat sidestick controller.

Town with fuel for $2\frac{1}{2}$ hours still on board despite having brought a load of 22 tons (22,353 kg) of spare parts and 50 people non-stop 10,290 miles (16,560 km) from Seattle. One of the latest versions, of 1982, is the 747-300 with a stretched upper deck which increases upper-deck seating from 32 to 91 and also raises the cruising speed because of the better contours. At the time of writing, almost 600 747s have flown 15 million hours and were carrying 5 million passengers each month.

It was natural for the new HBPR technology to be applied to giant wide-body transports designed for shorter sectors, such as those in Europe or US coast-to-coast. In Europe BAC, Hawker Siddeley, Sud and HBN (Hamburger-Breguet-Nord) all tried to get started but failed, while Rolls-Royce ran the RB.178 engine at 25,000 lb (11,340 kg) in 1966 but found most customers wanted a bigger engine. By 1967 the governments of France, West Germany and Britain had signed a protocol to build a new-technology 300-seater, the Airbus A300, with two RB.207 engines of 47,500 lb (21,546 kg) each. BAC tried to kill this off by offering the rival BAC.211 and then the 311, with engines hung at the rear. Eventually Rolls-Royce managed to sign up Lockheed as a customer for the RB.211 engine, rated at 33,600 lb (15,241 kg). The Lockheed L-1011 TriStar was a three-engined machine aimed in the first instance at US domestic routes, and it was launched in March 1968 by an announced order for 144: 50 for Eastern, 50 for a British group called Air Holdings to underwrite aircraft for export customers, and 44 for TWA. A few days later Delta signed for 28, giving a launch backlog of $2.58 billion, the best backing of any new transport in history. But McDonnell Douglas, which had an almost identical trijet in the DC-10, finally got orders from American and United and went ahead using the new CF6 engine from GE.

In Europe the A300 went ahead, but politics soon put sand in the works. There was no way Rolls-Royce could provide an engine for this big twin and also develop the smaller RB.211 for the new Lockheed L-1011 TriStar, and Airbus Industrie cut back the size to 250 seats in the A300B and went ahead with a more powerful version of the CF6, the CF6-50, initially rated at 49,500 lb (22,453 kg). A big international team was set up to build the Airbus, which was potentially a worldwide winner and more attractive to more airlines than the American trijets. But BEA in Britain scorned it, Lufthansa was critical and the British government pulled out entirely. Showing colossal cool nerve, Hawker Siddeley decided to stay in and finance the A300B wing with its own money. Eventually the first A300B flew on 28 October 1972. American rivals were initially successful in either damning the aircraft technically, calling it "the creation of a government committee that will never sell", or in spreading the impression that Airbus Industrie was an unknown and unreliable outfit that nobody dared buy from. In fact its members had been in business longer than most of the US builders, and under French law there was no way any one of them could escape from normal contractual obligations.

But for years hardly any A300Bs were

Right: Rolls-Royce went bust trying to build RB.211 engines for the Lockheed TriStar. Here a British Airways TriStar lands at Farnborough airshow.

sold, even after Air France began regular operations on 30 May 1974. Airlines appeared to find it hard to believe that a good airliner could come out of Europe, but perhaps the turning point was the loan of four A300B2 aircraft to Eastern in the USA. This giant and experienced carrier, perhaps to its surprise, found the European wide-body "the most efficient and least-troublesome airplane we have ever introduced to our routes" and it bought a big fleet. In the year 1979 alone Airbus received 133 firm orders and 88 options, including 12 new customers. Rather suddenly the world at last realised what any impartial observer could see from the start: the A300 is an absolute winner. The AI team went on to build versions with convertible passenger/cargo interior, long-haul models with transatlantic capability (which were politically barred at first, because twin-engined machines were judged unsafe until Boeing had a rival product, the 767) and the smaller and even more efficient A310, first flown on 3 April 1982.

As for the US trijets, McDonnell Douglas swiftly drew ahead of Lockheed because GE was able to offer a 50,000-lb CF6-50 engine, so that the preferred DC-10 became the heavy long-range DC-10-30, which sold all over the world. Rolls-Royce suffered so many problems with the RB.211 that none could be delivered, and all the company could do was fill warehouses at Derby with unsaleable parts. With money not coming in, the famous firm collapsed in February

First flown on 27 July 1972, the McDonnell Douglas F-15 Eagle is generally regarded as the natural inheritor of the title 'best fighter in the world' (outside the Soviet Union, at least) from its stablemate, the F-4 Phantom. These USAF F-15Cs are each carrying four Sparrow and four Sidewinder missiles, and the aperture for the gun can be seen in the root of the wing.

Above: Here seen carrying giant AA-6 missiles with the Libyan air force, the MiG-25 'Foxbat' was designed to intercept the B-70 (p 170).

1971, and was taken over by the British Government, but the takeover explicitly left out the RB.211. After a year of wrangling a new deal was struck with Lockheed, while bringing back from retirement such fine engineers as S. G. Hooker, A. C. Lovesey and A. A. Rubbra quite quickly turned the RB.211 into a saleable engine, and the TriStar at last entered service with Eastern in April 1972, almost a year after the DC-10. Both trijets were basically superb aircraft, but by a quirk of fate the DC-10 suffered a succession of highly publicised major accidents. Some were the fault of the aircraft but caused no casualties, as when a baggage door burst open, spewing out the contents of the cargo bay high over Michigan. Others killed all on board but were not the fault of the aircraft, as when an Air New Zealand DC-10 flew into Mt McMurdo in Antarctica. The two that really made headlines were a repetition of the baggage-door failure over the Forêt d'Ermenonville in France in 1974, when the passenger floor collapsed under pressurisation loads and all 346 passengers were killed, and an American Airlines DC-10-10 whose engine fell off the left wing on takeoff from Chicago on 25 May 1979, causing the slat on that wing to retract and the fully loaded aircraft to dive into the ground inverted. Mass-media articles suggested the DC-10 would (or should) never fly again, but the basically fine aircraft weathered the storm. Whereas Lockheed came to the end with the 250th TriStar, the DC-10 went on to No 367. Even today, more could be made, as the line at Long Beach is still building the KC-10A Extender tanker/transport version for the USAF.

Turning to military combat types, the 1970s saw a colossal build-up in the offensive power of Soviet air forces, largely using swing-wing (VG) aircraft inspired by the F-111. Throughout the decade the highest production rate was that of the MiG-23 and MiG-27 family of single-engined attack and all-weather fighter aircraft noted for their tremendous performance, versatile weapons loads and ability to disperse away from airfields in the way regularly practised by all Warsaw Pact air forces. Need for such dispersal appeared to be ignored by the NATO forces, all of whose extremely costly new combat aircraft continue to be sent from the factory to the only points in the world on which five or more missiles are continuously targeted! Meanwhile, the MiG (Mikoyan/Gurevich) bureau continued to update the neat tailed-delta MiG-21, which had first flown in prototype form in 1955. Early versions were extremely limited, but each year from 1958 improved models have appeared with more fuel, a more powerful engine, better and more varied armament and more comprehensive avionics. In the 1970s it was reluctantly realised that the basic design was nearing the end of its useful life, but with almost 10,000 built and production continuing

Right: Day-Glo red on this US Naval Air Test Center F-14A Tomcat makes it highly visible. Here it is parked aboard USS Forrestal, *with its swing wings folded back.*

614

15

116815
CANADA

Below: Even excluding Chinese J-7 production, about 10,000 MiG-21s have probably taken to the air since this MiG-21F was built in the early 1960s and supplied to Yugoslavia.

throughout the decade it posed severe problems to NATO on the score of numbers, if for no other reason.

The MiG bureau's big Mach-3 MiG-25 remained something of an enigma until in September 1976 Lt Viktor Belyenko defected from his base at Sakharovka and landed in Japan. Though in many respects the MiG-25 reflected its 1959 origin, and hardly any of its avionics were of the modern solid-state variety, it proved to be a tough and impressive machine which was minutely dissected, and the engines run, before being handed back. At no time was the original MiG-25 a fighter, in the sense that its airframe is designed for speed in a straight line rather than for close combat. Its existence, however, triggered off two very important fighter programmes in the USA which recognised the failure in that role of the F-111. In timing, the first was the Grumman F-14 Tomcat, the standard carrier-based fighter of the US Navy. Powered by two of the same old TF30 engines as used in the early F-111, it was nevertheless a superb swing-wing tandem-seater, with a very powerful radar and able to carry up to six Hughes Phoenix AAMs, the biggest and most powerful intercept missiles in the West with the ability to pick an aircraft out of a close formation and home in on it from a range of 100 miles (161 km). The F-14 also has an internal gun and can carry a wide range of other weapons for use against air and surface targets, but it proved so costly that, while the Navy insisted on renegotiation of the contract, Grumman said it had already incurred a loss of $105 million. Production came to a halt in 1973–74, but picked up on a new basis in 1975 and since then the only big problem has been prolonged difficulties with the engines.

McDonnell Douglas was luckier, in that its F-15 Eagle, the first contract for which was awarded in 1969, had new engines from the outset (though even these, Pratt & Whitney F100s, took a long time to

Pages 214–215: This CF-5D is the Canadian-built trainer version of the best-selling Northrop F-5 Freedom Fighter, from which the later F-5E Tiger II and F-20A Tigershark were developed. The inset picture shows six F-5E Tiger IIs of the USAF 57th Fighter Weapons Wing, each carrying one Sidewinder.

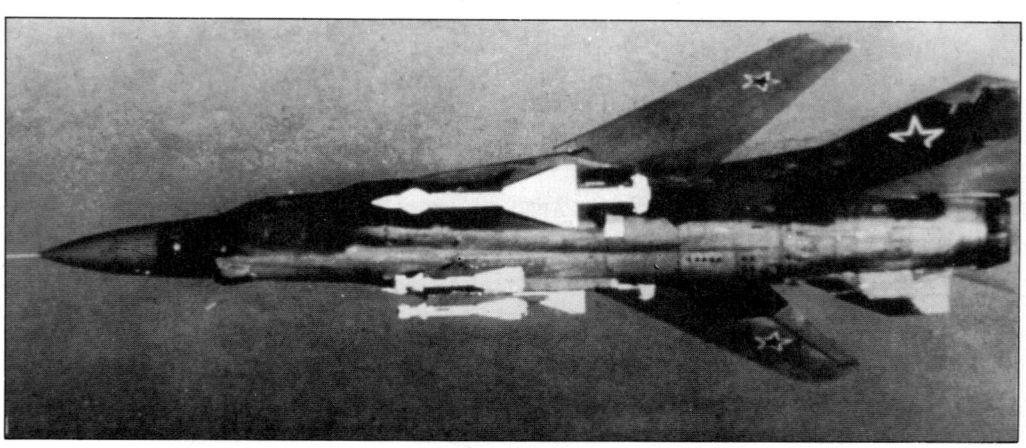

Right: A complete contrast to the MiG-21, the MiG-23 family have an engine of about twice the thrust and a high-mounted pivoted 'swing wing'.

mature to the reliability for which that supplier was always famous). By 1969 the VG wing had become rather less fashionable, and as the F-15 was designed to fly from long runways and engage in air combat, without bothering about the low-level attack mission, the large-area fixed wing was the best choice. It is a reflection on the supposed progress in aviation that the F-15, first flown in 1972, is a single-seat air-combat fighter yet is longer than any of the major twin-engined attack bombers of World War II, has a bigger wing area and more than twice the laden weight, and it burns fuel from eight to 12 times as fast! As for cost, this is anything from 300 to 1,200 times as great! At the same time, the fighting potential of the F-15 is impressive by any standard. For the first time the total installed thrust at low altitudes exceeded the clean gross weight, giving unprecedented agility in both the horizontal and vertical planes. Like the F-14 the new USAF fighter had two large engines with variable lateral inlets and twin vertical tails. When it entered service in late 1974 it introduced a completely new ability for the pilot to sit in state with a near-perfect all-round view. with a superb radar display

Below: The Mirage 50 differs from the earlier best-selling Mirage III family in having a more powerful version of the Atar engine.

In 1966 Dassault flew the first Mirage F1, a fighter totally redesigned from the Mirage III tailless deltas. Though Dassault has now returned to the tailless delta, with the Mirage 2000, the F1 became a best-seller in its own right. Here one fires a Matra Super 530 missile.

from which everything was erased except the key items he needed at any given time.

At the other end of the cost scale, Northrop continued throughout the 1970s to sell its neat F-5 supersonic fighters powered by two tiny GE J85 engines, and from the start in 1959 built not for the US services but for export. The F-5A and two-seat F-5B kept the production lines busy in the 1960s, and a few even served with the USAF in Vietnam to see how a simple lightweight machine would perform. In 1970 the US government held a competition to find a successor to these popular products, and Northrop won again with its F-5E Tiger II. This looks almost the same as the F-5A, though it has a rather more powerful J85 engine, a lightweight radar and various detail improvements. A few F-5Es were even bought by the USAF and Navy to act the part of hostile fighters, such as the MiG-21, in 'Aggressor' or 'Top Gun' training of American fighter pilots. Almost 1,000 of the original F-5 models were sold, but the F-5E passed this in 1982 and is well on the way to 1,500.

Almost the only other Western combat aircraft sold in such numbers are the French Mirages. The Dassault firm studied various replacements in 1963 and eventually decided on a different layout with a high-mounted, fixed non-delta wing and a rear tailplane, the result being the Mirage F1, first flown in 1966. An excellent all-round fighter, it entered service in 1973. It sold even faster than the delta, with 700 so far, but what Dassault did not expect was that alongside the F1 they would keep building the old Mirage III, the simplified Mirage 5 and the uprated Mirage 50. Indeed, in December 1982 the company flew the first Mirage 3NG, which is a modernised Mirage III for the 1980s! The completely different Mirage 2000 comes in the next chapter.

The only other European aircraft coming anywhere near Mirage numbers is the Tornado, launched in 1968 by Britain, West Germany and Italy in a most efficiently managed but very protracted programme to create what was originally called the MRCA (Multi-Role Combat Aircraft). One reason for the long timescale is that everything had to be agreed by the three governments, four customers (Germany has both the Luftwaffe and the naval Marineflieger), and the industrial groups Panavia (British Aerospace, MBB and Aeritalia) building the airframe and Turbo-Union (Rolls-Royce, MTU and Fiat) building the 16,000-lb (7,258-kg) thrust RB199 turbofan engine. The extremely compact engine, two of which are installed complete with variable inlets and nozzles and thrust reversers for short landings, is one reason why the Tornado has much greater capability than any other aircraft of its size ever built. With wings spread it can take off in less than 3,000 ft (900 m) with 10 tons (10,160 kg) of weapons selected from an unprecedented 112 different kinds of store, fly 863 miles (1,390 km), fold its wings back and attack at supersonic speed at a lower height than any other aircraft, and then return to base. Avionic equipment is of the highest possible standard, and the workload is shared by a crew of two in tandem; in a proportion of the 805 aircraft both cockpits are equipped for pilots, but usually the rear has special electronic displays for navigation, attack management and EW (electronic warfare). At the time of writing, 390 of these compact multi-role aircraft had been delivered, including the first of the longer and more rakish F.2 model used by the RAF as its next all-weather interceptor.

Below: Known as 'Backfire-B' to NATO, this swing-wing bomber is believed to be the Tu-22M. This one was carrying an AS-4 cruise missile.

To a considerable degree the Tornado is the successor to yesterday's bombers, though some nations still need large strategic aircraft. The Soviet Union has never stopped, and the lumbering Tu-16s, Tu-95s and M-4s, and supersonic Tu-22 ('Blinder'), were joined in about 1971 by the prototype of a very efficient and formidable VG swing-wing bomber known to NATO as 'Backfire', and believed to be the Tu-22M. Powered by two extremely large engines, it was derived from the Tu-22 but has new pivoted outer wings, giant engine inlet ducts and many other changes, including external racks for a heavy bomb load as well as pylons for various cruise missiles. Capable of a dash at Mach 2, it has enough range to cover much of the USA with flight refuelling, though it was really built for many other possible missions. As noted later, an even bigger bomber is coming along behind.

This bigger bomber looks like a scaled-up version of the USAF's Rockwell B-1. While the old B-52 has had to soldier on through the 1970s and 1980s, and now at least has new avionics and cruise missiles to increase its combat effectiveness, its basic age and enormous reflection when seen on hostile radars has made its replacement a matter of urgency for the last 20 years. Accordingly, after an unprecedented 10 years of study – so that one of the proposed bombers, the AMSA (Advanced Manned Strategic Aircraft) was said to mean 'America's Most Studied Airplane' – Rockwell was at last given a contract to build a new swing-wing bomber in 1970. Powered by four new GE F101 turbofan engines, the first B-1 flew on 23 December 1974, and was at that time by far the most advanced instrument of long-range airpower ever built. Four prototypes were built, but in 1977 President Carter, rather strangely presenting the cruise missile as a new invention that, allied to the B-52, would render the B-1 unnecessary, cancelled the project. In October 1981 it was resurrected, with various improvements, and the first of 100 B-1Bs is expected to join SAC in 1985.

It is naturally difficult for the ordinary air enthusiast to feel affinity with such a costly, deadly and complex machine. Yet world-wide interest in aviation has never ceased to grow, and in the 1970s the

Below: This RAF Tornado is seen at low level with eight 1,000-lb (454-kg) bombs. In 1984 RAF Tornados took the top trophies in a major contest against the USAF.

FIESTA

NAVY
an adventure
N9492L

Previous pages: As recently as 1960 there were hardly any hot-air balloons in existence. Almost 200 years after its invention, it staged an amazing comeback

enthusiast movement multiplied in a spectacular manner. The number of powered private pilot licences grew only modestly, though those endorsed for helicopters doubled. But the number of hot-air balloons quadrupled, and the number of homebuilt aircraft grew by an even larger ratio, though as many constructors are taking many years to build their first aircraft precise figures are hard to calculate. A more precise figure is the knowledge that in the 1970s more small engines for homebuilts and microlights were sold than in the whole previous history of aviation.

At the same time, the 1970s saw an approximate doubling in the number of aviation museums around the world, and probably a great increase in the number of passionately concerned people who tear their hair out at the historic and unique aircraft thoughtlessly bulldozed into fragments since 1945. Those aircraft that are left are now lovingly preserved, though this can give future generations a distorted idea of history. For example, a recent major article on the P-51 Mustang fighter of World War II was illustrated with 15 colour photographs, all showing the P-51D. In fact this model only came into use in 1944, and the Allison-engined Mustangs that fought through all the toughest days of that conflict have been almost forgotten, largely because until 1983 not one example existed. Today, a single specimen has been restored and flown in RAF markings.

It has been estimated that the number of people actively engaged in restoring old aircraft is in excess of 45,000, or considerably more than the strength of most air forces. The problem is not one of manpower, or dedication, or money, or the will to do the job, but simply lack of knowledge, lack of original drawings and lack of genuine parts. Almost all the pristine squadrons of famous old aircraft have wheels, propellers, radios and even engines that are not the same as the original. Hurricanes have been restored with six-stub ejector exhausts never seen on the original. And a British team has been thwarted in its wish to fit a front gun turret back on the only surviving Wellington because nobody can discover how the turret was mounted. Yet the number of Wellingtons built was 11,461!

Right: The USAF hopes to buy 100 Rockwell B-1B bombers at a cost of $20,500 million,

McDonnell Douglas F/A–18s of the US Navy and Marine Corps armed with two AIM-9 sidewinders mounted on wing-tip rails.

10. FULL CIRCLE

It was inevitable that aeroplanes should progressively diverge into two major groups, the incredibly complex and expensive for the professional operators and the simple and affordable for the private owner. What might not have been predicted 50 years ago, for example, is the great worldwide fleet of intermediate-technology General Aviation machines such as executive jets, helicopters and crop-sprayers, and even as recently as the late 1960s it is doubtful that anyone could have foreseen the tremendous boom in hang gliders and microlights.

Getting the common man into the sky was the basic objective of the earliest aviators, few of whom can have had many thoughts about combat aircraft or giant airliners. In the early 1920s the British twice held official design competitions for light aircraft, but they came out so heavy and underpowered as to be of little use, and it was left to Capt de Havilland to produce the Cirrus engine and the two-seat Moth. In the 1930s Henri Mignet in France wrote a book about how to build a *Pou du Ciel* (Flying Flea); it became a best-seller and hundreds of Fleas were made in many countries. Again, the idea was better than the actual design, and many young enthusiasts killed themselves trying to fly an ill-conceived machine.

Today we have probably got the balance right, and sporting aviation in most countries is properly controlled by experienced yet positively minded bodies who keep the casualties to a minimum. Some are concerned with obviously dangerous sports such as racing, aerobatics and parachuting, while the vast international EAA (Experimental Aircraft Association) does a fantastic job in encouraging the home-builder whilst steering him (or her) away from faulty design or manufacture. Home-builts are pure fun aircraft, some being for speed, a few for aerobatics, and include in their numbers amphibians, autogyros and helicopters. They embrace every possible form of configuration, construction and materials, and so far a significant proportion have been replicas of famous wartime fighters, from both world wars, redesigned on a reduced scale with low-power engines.

Modern competition sailplanes are perhaps the most beautiful of all aircraft. They are organised broadly into a Standard Class, with a wingspan not exceeding 49 ft 2½ in (15 m), and an Open Class, with unlimited span and if necessary such additions as flaps, airbrakes, water ballast, a tail parachute and a computer and other instrumentation for continuously searching for the best atmospheric lift. The best Open machines are pure poetry in glassfibre and carbonfibre, with incredibly elongated (the technical term is high aspect-ratio) wings, and able to travel 55 to 60 metres forwards for every metre of height lost. The self-launching or motorised glider is basically an established glider type fitted with a small retractable engine/propeller group in the top of the fuselage, or with a conventional engine/propeller installation in the nose. It is not a new idea, but it has

Below: Though it has a 'narrow body' the Boeing 757 has proved as popular with passengers as it has with airline accountants. Most, including this one, have Rolls-Royce 535 engines.

remained a rarity until the 1970s. Today hundreds are in use and they are becoming more popular all the time. A small number even have tiny turbojets, an example being the Italian Caproni Vizzola A-21SJ, which seats two people side-by-side and can fly them 217 miles (350 km) in well under two hours.

Microlights and hang gliders are a recent breed of 'minimum aircraft', with empty weights varying from about 50 lb (23 kg) upwards. They come in assorted shapes, a popular configuration having the pilot, and whatever airframe is provided, hanging under a delta-shaped flexible wing of Rogallo type, with no separate tail. Other arrangements include swept wings, canards and biplanes. In almost all cases the airframe, such as it is, is of welded aluminium tubing with a Dacron fabric covering. The Microlights, which have engines (typically of about 18 to 45 hp), take off at around 25–30 mph (40–48 km/h) and cruise at perhaps twice this speed, the one or two occupants usually sitting inside some form of streamlined nacelle giving protection from the wind, though in about one-third of current machines they are completely in the open. In almost all cases there is a proper three-axis control system, whereas in the hang gliders the commonest form of control is by the suspended pilot shifting his weight, just as Lilienthal did in the previous century.

A highly specialised class are the man-powered aircraft (MPA). Building an aeroplane (or should it be considered a glider, as it has no engine?) powered only by its occupant(s) has engaged the attention of many experimenters since 1930, but the task proved too difficult to accomplish. In the 1950s a British industrialist, Henry Kremer, offered a prize which eventually grew to £50,000 for an MPA that could be flown around a figure-8 course around two pylons half a mile apart. A few misguided people tried to build MP helicopters and ornithopters (flapping-wing machines), but the obvious way to win was to build the lightest possible wing of the greatest possible span, typically around 100 ft (30 m), and use it to support the smallest possible airframe and tail, with a saddle and pedal drive to a wheel and also to a very large slow-turning propeller. Dozens of groups failed to win the Kremer Prize, but a California team headed by Dr Paul B. MacCready built Gossamer Condor. This had a canard foreplane, was built of balsa tubing, corrugated cardboard and foam plastics, braced by fine steel cable and covered in Mylar film. On 23 August 1977 Bryan Allen pedalled round the course in just over six minutes, averaging 10.8 mph

Above: Harrier GR.3 attack/recon aircraft of RAF No 3 Sqn. Only a few have been sold as it seems that most air forces ignore the vulnerability of airfields.

(17.4 km/h). Next MacCready built Gossamer Albatross, a refined model with carbon-fibre structure and fewer external wires. On 12 June 1979 Allen boldly pedalled away from near Folkestone, England, and nearly three hours later landed at Cap Gris Nez, France, after having for some time encountered headwinds. His airspeed averaged about 14 mph (22.5 km/h). MacCready's team then went on to build Solar Challenger, with solar cells in the wing powering a small electric motor driving a tractor propeller. This machine, with a relatively short span because of its higher speed, made many fine flights, including one of 8 hours 19 minutes, but its most famous achievement was the flight of 7 July 1981 from Pontoise, north of Paris, to Manston, Kent, a distance of 229 miles (368 km), flown at an average of 42.5 mph (68.6 km/h). MacCready commented, however, that solar cells are too costly for such machines to be of much 'practical or commercial use'.

Below: *Bert Rutan designed lightplanes of amazing flair and efficiency. In 1984 over 2,500 of these Long EZ two-seaters were being built by home constructors.*

A very different idea that so far has failed to be of commercial use is jet V/STOL. From 1960 such projects appeared frequently, in order to carry passengers at jet speed between city centres and to avoid the queueing and overcrowding of air traffic at airports. A typical proposal was Hawker Siddeley's HS.141, a 600-mph (966-km/h) 119-seater with little out of the ordinary except for large bulges along the lower sides of the fuselage, ahead of and behind the wing and extending almost the whole length of the aircraft. These were to house a total of 16 RB202 lift fans, highly specialised turbofan engines giving about 12 lb (5.5 kg) of lift for each pound of weight. These could lift the machine vertically, though the normal method would have been to use a short runway. Noise was not regarded as an insuperable problem, but the basic economics and hazards of such a programme combined with the soaring price of fuel to kill it.

After carefully studying alternative short-haulers the same team finally settled on the HS.146, a much more conventional STOL high-wing machine using four of the extremely quiet and economical Avco Lycoming ALF502 turbofans, each of 6,700 lb (3,040 kg) thrust. Though a small aircraft weighing a mere 76,000 lb (34,473 kg) and with 70 to 90 seats, the cabin was as wide as a 707 and wider than a DC-9, and passenger appeal was rated as exceptional. The wing was extremely small, but its extremely powerful flaps put the 146 in the true STOL class, able to use unpaved strips less than 3,300 ft (1 km) in length. It was calculated that it would be the most agile and easily flown airliner in history, yet at the same time the quietest. All was set for a 1973 go-ahead, but soaring

Right: *Another Rutan design is the Solitaire, a self-launching sailplane. It has a glassfibre airframe and a retractable 23-hp engine.*

Bottom right: *With the Rolls-Royce 535E4 engine the Boeing 757 performs about 75 per cent more passenger-miles per gallon than the best previous medium-range jetliners.*

inflation, the fuel crisis, industrial unrest and poor marketing prospects combined with nationalisation of the British industry to force a halt. A small team kept refining the design, and eventually the newly formed British Aerospace launched the programme a second time, as the BAe 146. The first one flew on 3 September 1981. Fokker of the Netherlands appeared to be incensed that Britain should build such an aircraft, because it would compete with its own F28 Fellowship, and there was a surprising amount of criticism of the 146 based on such features as its four engines, which for unexplained reasons were thought to mean high costs. Britain had not built a new transport for 20 years, and with the 146 great care was taken to get the specification exactly right and the engineering design better than in any competitor aircraft. Just like the European Airbus, so did this British machine have a slow start and an uphill struggle against closed minds who apparently refused to believe that a British machine could sell in the marketplace. Of course, once airlines (and military customers) began to try out the 146 the orders at last began to roll in, and there is no doubt that by the end of the decade the 500 mark will have been passed.

Boeing, on the other hand, has such a reputation that some civil customers, notably Ansett of Australia, made a point of refusing to buy the A300B but of waiting until a duplicate had been made by the US firm. The 767, first flown on 26 September 1981, is indeed almost a Chinese copy of the European machine except for having a narrower body and a bigger wing. The reason for the latter was that some airlines

Below: First flown in August 1947, the Antonov An-2 appeared outdated, because it was a biplane. In fact, production has exceeded that for any other aircraft type.

Right: 85 per cent full-size, this prototype of the Beechcraft Starship 1 shows the unusual layout of this 400-mph (644-km/h) twin-turboprop.

wanted the ability to climb straight up to about 40,000 ft (12.2 km), but the narrower body was difficult to appreciate. Boeing and their customers have always explained that the big advantage of the US machine over the A300B is that, as it has only three instead of four seats between the aisles, every passenger can have an aisle or window seat except for the final 15 per cent, overlooking the fact that the A300B can

Above: The cockpit of the Starship 1 has the uncluttered look of all modern flight decks in which individual 'instruments' are replaced by instantly reprogrammable displays.

Overleaf: Every airport in the Western world has a GA (general aviation) area where private and business aircraft can park, with the fullest facilities.

31
N1996G

have the same seating arrangement with even more spaciousness! An immediate practical effect of the smaller body is that standard LD3 baggage or cargo containers do not fit, and Boeing had to produce a non-standard smaller container called the LD2.

Timed only a few weeks later, Boeing also launched a new narrow-body to succeed the 727, designated the 757. At first this was to incorporate many 727 parts, including the complete nose and cockpit, tail and landing gears. After years of refinement it matured as a completely new aircraft, though with the same 140 in (3.55 m) cabin width and basically the same fuselage, though much longer. Two completely new advanced-technology engines were hung on the almost unswept wing, and for the first time a US trunk-route transport was launched with British engines, two Rolls-Royce 535s of 37,400 lb (16,960 kg) thrust. These promised unrivalled low fuel consumption per passenger, but Pratt & Whitney was forced to offer competition and at great expense has produced the PW2037 in the same class. The first 757 flew on 19 February 1982, and quickly showed itself to be a real winner in the style of the 727, with many per-

Below: This picture of the X-29 on vibration test prior to first flight gives just a hint of why aerospace is an expensive business. (The soft toy is not charged to the customer!)

Right: Grumman's X-29 has been built to explore the FSW (forward-swept wing), which until recently was structurally unattainable.

GRUMMAN

103

CCV
29-5103
TR01

Previous pages: This CCV (control-configured vehicle) research aircraft is a Japanese Mitsubishi T-2 supersonic trainer.

formance figures even better than prediction. Likewise Rolls-Royce also turned up trumps with the 535, producing not only an engine with better than predicted performance but also one setting a new standard of reliability. In January 1984 it was announced that in the first year of airline service the 535 had had to be removed from the aircraft only twice in 78,500 hours; this was "more than ten times the previously claimed industry best, and equivalent to nearly 20 million miles of flying between removals". Whose car engine can say as much? In 1984 the even better 535E4 entered service, with an extra fuel saving of 15 per cent over the original model as well as higher thrust. All indications were that this would set even better records of reliability.

Rather surprisingly, the extremely capable Soviet designers have consistently run into technical difficulty with all their big civil transports, and have had to spend as long as ten years getting them right. This has certainly been the case with the Il-62, the first Soviet long-haul civil jet, and with the smaller 120-seat Yak-42, of which as many as 2,000 may be needed. By 1981 the Yak-42 was in full service and it is planned to become the standard Soviet vehicle except on long trunk routes. For the busiest routes the Ilyushin bureau produced the big Il-86, looking like a 707 but as big as a DC-10. This follows Soviet practice in that passengers carry their baggage on board, enter through any of three airstairs built in on the left side and walk up to the lower deck, where they stow their cases. Three interior stairways then lead up to the main deck with its 350 seats. Rather unusually for a Soviet civil programme, the Il-86 is assembled at Voronezh from parts made in widely separated factories, the fin, tailplane, ailerons, flaps, slats and engine pylons being assigned to the Mielec works in Poland.

Normally Poland is tasked only with General Aviation aircraft for the Soviet bloc. Though it has lately received stiff competition in this field from a surprisingly independent Romania, it had for many years had a corner in the vast market for agricultural machines, of which thousands are

Right: The F-16 AFTI (Advanced Fighter Technology Integration) has two small fins projecting down from under the air inlet. The whole aircraft has been replanned to perform 'impossible' manoeuvres.

UNITED

Previous pages: Though it resembles the obsolescent Mirage III the Mirage 2000 is a totally new aircraft embodying the latest technology. This example was demonstrated with eight 551-lb (250-kg) bombs, two 374-gal (1,700-l) tanks and two Magic missiles.

Page 243: United is the biggest airline apart from the Soviet Aeroflot. It has 50 of these DC-10s, but these are outnumbered by 240 Boeings.

Right: The Boeing E-3 Sentry fits into the airframe of the 707-320B transport an extremely capable radar surveillance and communications system.

used in the Soviet Union. All the modern ones are angular low-wing monoplanes with the pilot perched high for good all-round view, but the most numerous in the Soviet Union is still the An-2 biplane. This first flew on 31 August 1947, and Western observers may have been forgiven for thinking that a biplane was a quaint idea at that time. In fact the An-2 was a most useful machine with a modern stressed-skin structure, a fuselage like a short DC-3 and a 1,000-hp engine able to pull it out of impossibly short strips (takeoff run on a runway is 492 ft/150 m!). Far from being obsolete, almost 5,500 were built by 1960, by which time manufacture had been transferred to Poland. Here more than 9,200 had been built by the end of 1981, with probably another 2,000 due by the completion date in 1985. In China a further large number, not yet known but probably about 3,000, have been built as the Y-5, with production continuing. The total of almost 20,000 is certainly greater than that of any type of monoplane since 1945!

Of course, if one takes all the high-wing four-seat Cessnas as a single group – though they cover a multitude of different models – one gets numbers which today seem astronomic. Cessna of Wichita, Kansas, has delivered well over 35,000 of the Model 172 and Skyhawk series, as well as well over 20,000 in the Model 182 and Skylane series, and altogether by 1984 Cessna had delivered 175,000 aircraft. The only company that can rival it for numbers is Piper, which has moved its headquarters to Vero Beach, Florida. All the first Pipers were simple high-wing machines, but for many years the main Piper output has concentrated on attractive low-wing cabin models ranging up to the Cheyenne IV twin-turboprop eight-seater which can cruise at 404 mph (650 km/h). Piper delivered its 100,000th aircraft in 1976 and had reached 132,000 by 1984.

The third of the 'big three' builders of lightplanes, Beech, had yet to reach 50,000 by 1984, though it was very close. In 1983, while continuing to market such established twin-turboprops as the Super King Air, Beech flew the prototype of a new 10/12-passenger machine with a radical canard configuration proposed by Mr Bert Rutan. He has produced a series of unconventional lightplanes which have not only looked good but have demonstrated high

Below: The Nimrod AEW.3.

Below: Sweden's main air defence fighter is the Saab Viggen. It is seen here armed with two Hughes' Maverick air-to-ground missiles.

performance and extreme economy. Moving up into the executive transport field was a natural, and the Beech Starship 1 promises to be a world-beating machine with a cruising speed of 400 mph (644 km/h) resulting from two pusher engines of 1,000 hp each mounted in pods above the rear inboard sections of the swept wing. Yet another famed modern designer was Bill Lear, who launched the best-selling Learjet family. These business jets were the E-types of that market, because they were made sleek for speed and good looks, at the cost of not enough headroom to stand up inside. The argument is, if you get there fast enough you can stay in your seat! Lear died in 1978, leaving designs for another radically unconventional yet impressive light transport, the Learfan. This is one of the most streamlined aircraft ever built, with a pressurised fuselage for up to nine passengers and a pusher propeller of special low-noise form right at the tail, driven by two 850-hp engines in the rear fuselage. The engines in both the Learfan, which is made in Northern Ireland and can cruise at 418 mph (673 km/h), and the Starship are PT6 turboprops, adding further numbers to the colossal total already made of the engine the Quebec officials refused to help launch, as noted earlier.

The trend towards unconventional layouts is not due to a mere wish to be different, because both the Starship and Learfan are programmes involving a risk of about half a billion dollars, and the final product simply has to be better than the competition. In the vast field of 'homebuilders' the designer can often indulge in his own eccentricities and build for fun – so long as the machine is safe to fly – but the record shows that, after an amazingly long delay of well over 60 years, designers are at last realising that perhaps the traditional aeroplane is not always the best and that a propeller at the tail, or a tail at the front, may be superior.

The move towards canards, with the horizontal tail at the front, has been especially strong in air-combat fighters. In the 1950s the choices facing the designer were the swept wing, the delta and, for purely supersonic machines, the small razor-thin straight wing, as on the F-104. By 1960 the variable geometry swing wing was all the rage, and by 1970 the preferred arrangement was the ordinary fixed wing with virtually no sweep other than taper on the leading edge. By this time designers were using electrically signalled powered flight controls with such instant response that they were trying a new technology called CCV (control-configured vehicle), and by the 1980s this was beginning to come into general use with such new fighters as the Mirage 2000 and F-16. The CCV concept is the reverse of the usual one; most aeroplanes are made positively stable, so that they keep flying the way they are pointing. The CCV fighter is made deliberately unstable, so that it keeps wanting to swap ends and fly tail-first (which if done at high speed would be catastrophic). The control system continuously senses this tendency and keeps feeding in a stream of corrections to keep it flying nose-first. The process has been described as just like a cyclist sitting on the front of a car and holding the handlebars of a bike pushed back-end first at 60 mph. Our cyclist would lose control instantly, but with a CCV control system would keep the unstable bike pointing straight ahead.

Of course, one of the several advantages of CCV is that, when the pilot wants to manoeuvre, the fighter responds instantly and with lightning speed. Some early jet fighters were big and sluggish, but today's can beat even the old biplanes (although, because they fly much faster, they need a much greater radius of turn). A second advantage of today's fighters is tremendous engine power, so that they can fly straight upwards if necessary whilst accelerating through the sound barrier, and manoeuvre as well in the vertical plane as in the horizontal one. They also need large wings, in order to make a sustained turn at a bone-crushing acceleration quite beyond anything previously possible.

For example, the F-16 Fighting Falcon, built by General Dynamics, can hold a turn

Main picture: Today's RAF relies very heavily on inflight refuelling, using the probe/drogue method. Here a Victor K.2 tanker (a converted bomber of the 1960s) replenishes a Nimrod ocean patrol and anti-submarine aircraft.

Inset: The Boeing E-3 Sentry AWACS (Airborne Warning And Control System) provides surveillance over sensitive areas of the Western world. Seated at their consoles are men of the 961st Sqn.

Below: The Shorts 360, a 36-seater built in Belfast, has proved a best-seller among today's burgeoning twin-turboprop regional airliners.

at a sustained acceleration of 9g, which is almost twice the limit of piston-engined fighters and early jets. It is impossible for readers to imagine what such a load is like, though fighter pilots who have experienced 6g may get some idea of it. To meet it the pilot needs a good anti-g suit and to lie back in an inclined seat with his feet high on the pedals. The F-16 cockpit was specially configured for such manoeuvres, and also offers an almost perfect all-round view, with no frame panels anywhere. This superb fighter was one of two built in 1974 to test the capability of an LWF (lightweight fighter). Both the F-16 and the rival Northrop YF-17 were so good that pilots became excited and pressure grew for one to be bought for the USAF. Eventually the F-16 was chosen, powered by a single F100 engine of the kind already used in the big twin-engined F-15. Four NATO European nations – Belgium, Denmark, the Netherlands and Norway – picked it, and helped to build 1,000 by 1984. General Dynamics developed an enlarged tailless model, the F-16E, with a bigger engine and even greater manoeuvrability and weapon load.

Meanwhile Northrop's YF-17 had been developed in partnership with McDonnell Douglas into the F-18 Hornet for the US Navy. Powered by two 16,000-lb (7,258-kg) thrust GE F404 engines, this was the first aircraft designed from the start to be equally good at air combat and ground or anti-ship attack, without modification for either task. Fully equipped for carrier operation, the Hornet has also been sold to land-based air forces such as those of Canada, Australia and Spain, largely on account of its ability to effect stand-off all-weather interceptions with radar-guided Sparrow missiles, which could not be guided by the rival early versions of F-16. Perhaps the only adverse features of the Hornet are its large long-span wing, which is the opposite of what is needed for the low-level attack mission, and the fact that, planned as a 'cheap' replacement for the F-14, it escalated in price until it is even more expensive than the larger fighter.

It has already been noted that an example of the CCV formula is the French Dassault-Breguet Mirage 2000. At first glance this looks very like a Mirage III of 1956, but in fact it is a totally new design with more sophisticated structure, materials, systems, avionics and weapons. Thanks to a fully variable-camber wing, with hinged leading edges, the trailing edge comprises elevons which can assist tight turns and also help slow the landing (in the Mirage III the elevons had to be raised for landing, in effect thrusting not upwards but downwards). Despite limited French funding the Mirage 2000 has been developed as an extremely good interceptor and air-combat fighter, marred only by delay to its radar, and it has sold well in export markets despite a very high price. Future versions include tandem trainers and a special low-level nuclear strike variant able to carry the ASMP long-range attack missile.

Sweden, maintaining strict neutrality, has completed development of the JA37 fighter version of the Saab Viggen, the outstanding short-field combat aircraft which has demonstrated its ability to operate from small farm fields, straight sections of highway (closed to the public) and even country dirt roads. Sweden, more than most air forces outside the Communist bloc, realises that in a war the first targets to be destroyed would be the airfields. The next-generation JAS39 Grypen multi-role

combat aircraft will likewise be able to operate away from bases of known location. So too, in all probability, will be the very similar Israeli IAI Lavi (lion cub). It seems crazy that Sweden and Israel should spend so much developing two similar aircraft for the same job, but politically collaboration between them was not possible.

Panavia, comprised of companies from Britain, West Germany and Italy, has now completed the outstanding Tornado programme, with 805 production aircraft to which will be added four refurbished pre-production machines as well. Most of the final batch are of the F.2 interceptor version, 165 of which are entering service with the RAF. A purely British development, the F.2 is certainly the world's most efficient long-range all-weather interceptor, with the ability to patrol for 2½ hours at a distance of well over 350 miles (570 km) from base carrying a 27-mm gun, four Sky Flash and two Sidewinders. In RAF service it forms the air-defence partner to the airborne warning and control system aircraft. The original AWACS was based on the airframe of the Boeing 707 Intercontinental, and has a large radar installation with a 30 ft (9.14-m) rotodome aerial mounted on a pylon above the rear fuselage. In such a position the radar has imperfect vision, but the sheer power and sophistication of the Westinghouse radar in the Boeing E-3 Sentry series makes it a giant asset to any air force, able to see and direct almost all aerial activity within a radius of some 240 miles (386 km). Special communications links can be set up with 98,000 friendly bases without an enemy being able to interrupt or listen-in. The USAF has 24, to which up to ten more may be added, NATO has 18 bought under a cost-sharing deal and usually based in West Germany, and Saudi Arabia has also bought a version of these incredibly costly machines. The Nimrod AEW.3 project, with its aerials at nose and tail each covering a 180° sector, has perfect radar vision. The RAF ordered 11 of these before the British government abandoned the idea and opted for the American AWACS.

The unexpected war over the Falkland Islands in spring 1982 drove home to the RAF that to sustain a conflict 8,000 miles

Below: Despite its cost the Beechcraft Super King Air 200 series has sold all over the world, the total by 1984 reaching 1,500. It can carry 14 passengers long distances.

Left: An historic photograph taken on 20 July 1969 by Neil Armstrong during the first manned visit to the Moon. Ed Aldrin has deployed the passive seismic experiment package.

Below: Spaceflight entered a new age on 12 April 1981 when Shuttle OV-102 Columbia *lifted off on its first test comprising 36 Earth orbits.*

Left: Engineers at Pratt & Whitney Canada ready with the new PW120 engine for testing. Destined for airliners of the 1990s, it drives a propeller.

from Britain and 4,000 miles from any friendly base it must rely greatly on inflight refuelling. Urgent modification programmes were put in hand to equip Nimrods, Hercules and Vulcans with probes, and the Hercules with an HDU (hose-drum unit) to serve as a STOL tanker. British Aerospace has meanwhile been rebuilding a fleet of ex-airline VC10s and Super VC10s as long-range tankers, and Marshall of Cambridge has done the same with ex-airline TriStars. In former days new engines used to be subjected to years of tough military use before being sold to airlines, but today the financially extended RAF has at last received the RB211 engine secondhand after years with British Airways!

Two classes of aircraft which appear somewhat overcrowded, as far as the number of different types is concerned, are propeller and jet trainers and turboprop local-service airliners. There are more than 45 different species of the former and at least 31 of the latter, all fighting furiously for a piece of what is, at best, a limited market. In more than half the total (nearly all the small transports) the engine chosen is a turboprop in the 600 to 1,100-hp class. A few of the new transports have advanced technology turboprops in the 2,000-hp bracket, at last taking over where the Rolls-Royce Dart left off after a production run which, at the time of writing, exceeds 38 years. The small twin-turboprop airliners come with high or low wings, square unpressurised fuselages (and big windows) or circular pressurised ones, sleek civil styling or severe military lines with a rear loading ramp, and even with fixed landing gear in many of the high-wing machines. Different customers put different values on speed, comfort, payload, freight capability and operating cost, but all need reliability and environmental acceptance, which mostly means low noise.

Since the early 1970s various agencies have been studying the propfan, a multi-bladed propeller almost indistinguishable from the large-bladed fans used on the very latest turbofan engines. The blades are very thin, often scimitar shaped, and are able to operate at supersonic speed. Thus, the aircraft to which they are attached can be pulled at jet speeds, of perhaps 550 mph (885 km/h), but with much lower fuel consumption than for a jet. The only real problem has been severe noise, and this affects not so much people living near the airports as the passengers inside the aircraft. Noise is often difficult to overcome, but an obvious way to leave most of it literally behind is to put the propfans at the tail, either with a large one on the rear tip of the fuselage or with two on pods on the tips of the tailplane; alternatively, with a canard configuration there is no problem, the answer coming out like the Beech Starship. McDonnell Douglas was one of the first to publish an artist's impression of a possible propfan machine, a kind of con-version of the DC-9, but in this case the propfans were in line with the passenger cabin and thus unsuitably placed. It remains to be seen how far the propfan can be brought off the drawing board and on to the air routes but the prize is, loosely expressed, jet speed with turboprop fuel consumption.

Future historians may think that the greatest advance in the 1980s in the whole field of flying – in the broadest meaning of the word – was the US Space Transportation System, better known as the Shuttle. The practical space age began with the launch by the Soviet Union of a simple satellite called Sputnik 1 on 4 October 1957. Since then thousands of objects have been launched into space, most into Earth orbit but some into orbit round the Sun and a few to travel to the Moon or planets, and a very few have escaped from our Solar System and may go on for ever in the vastness of the Universe. Until the 1980s the launch vehicle had to be a giant rocket, mounted vertically and thrust upwards by its engines alone. Wingless, the mighty rocket would fall to Earth when expended, the launch usually being arranged so that it would crash into the deep ocean and sink. This naturally made space travel expensive; imagine the price of an air ticket if a 747 was thrown away on each trip!

With the Space Transportation System the NASA (US National Aeronautics and Space Administration) at last realised what hundreds of workers had envisioned for many years: making space travel more like air travel, using a vehicle that could be landed back on Earth to be used again and again. The Shuttle Orbiter, the main part of this system, was developed chiefly by Rockwell, which had earlier played a central role in the Apollo programme to put humans on the Moon and bring them back. Each orbiter is a tailless delta aircraft about as big as a 100-seat twin-jet, but with a small blunt wing, stumpy fuselage and a large group of rocket engines at the tail (which looks unstreamlined). Like other space launchers the Orbiter takes off pointing straight up, boosted not only by its own engines, which are fed with propellant liquids from a giant external tank, but also by two colossal solid rocket motors each of which weighs about 570 tons (579,156 kg) and thrusts upwards with a force of almost 1,200 tons (1,219,276 kg) for two minutes. These motors and the tank are then dropped off, and the motors are recovered at sea, towed back to land and used again. The Orbiter continues into space under the control of its crew on the airline-style flight deck, and it can be used to deliver or collect satellites, bring materials for a space station or do many other tasks. The first Orbiter, OV-102 *Columbia*, made its first flight in April 1981. It has since flown many times, and in April 1983 it was joined by OV-099 *Challenger* and then by OV-103 *Discovery*, which both flew. OV-104 *Atlantis* was completed but, along with *Discovery*, was grounded following the disastrous flight of *Challenger* in January 1986, which killed all the crew. NASA is hoping to resume operations during 1989.

To close on a more down-to-earth note, perhaps the other big development of the later 1980s will be the long-awaited rebirth of the airship. The image left behind by the old airships of structural failure, fire and large-scale disaster was due to the poor technology of the day, rather than to any fundamental shortcomings. Since 1965 many enthusiasts – at first looked upon as cranks and near-lunatics, as were the first aviators – have been proposing various updated airships for various purposes. Some have envisaged gigantic vehicles with carbon-fibre envelopes, large turboprop engines and a payload rivalling that of a modern container ship. Others have set their sights nearer to what the available funding can allow, in the form of almost traditional non-rigid blimps of modest size. One of the more successful is the British Airship Industries Skyship 500 which in 1984 appeared likely to be built in substantial numbers for several military and civil customers including the US Coast Guard and British MoD. The bigger Skyship 600 had been studied for many tasks before its first flight in 1984, including support of radar stations in the far Arctic, ocean surveillance and long-duration missions as an 'eye in the sky' surveillance aircraft.

For very short missions there is nothing to beat the sheer utility of the helicopter, the technology of which has now been taken by the Soviet Union into the 23,000-hp class with the Mil Mi-26, a most useful and versatile vehicle especially in the opening up of the vast, unexplored Siberian territories.

For pure observation missions an aeroplane can be more fuel-efficient, and here the British Edgley Optica has the market all to itself. First flown in December 1979, the Optica is essentially a transparent bubble riding in front of a long-span pusher aeroplane designed for 13 hours' endurance rather than for speed. But for the long on-station kind of surveillance there seems to be no answer to the modern airship, especially one with full avionic equipment and sufficient speed to hold station in a full gale. Just how many airships will actually be built in the remainder of the century is anyone's guess. Because they are quiet and theoretically efficient the apparent rebirth of lighter-than-air flying of a serious nature, using airships, could well become as impressive a growth as the ubiquitous hot-air balloon. Montgolfier would have been delighted.

Right: To save fuel propellers are also found on the latest trainers. This Pilatus PC-7 serves an airline, but most fly with air forces.